AAPG REPRINT SERIES No. 21

Fracture-Controlled Production

Compiled by
JOHN R. KOSTURA
and **JOHN H. RAVENSCROFT**

Published by The American Association of Petroleum Geologists
Tulsa, Oklahoma, U.S.A., 1977

preface:

Fracture Controlled Production:

To meet the increasing national demand for energy, petroleum geologists are turning their attention to subtle trapping mechanisms and more efficient secondary recovery programs. The role of fracturing in creating and enhancing reservoirs is one of the more elusive and challenging geologic problems facing petroleum geologists.

In reviewing papers related to fracturing, the diversity of problems, approach, reservoir characteristics, scale of investigation (from microscopic investigation to aerial photography), theoretical discussion, experimental technique, and production or lack of production induced by fracturing is impressive. It was not a simple task to select papers for this reprint volume as each paper has its own character and intrinsic value. Articles selected for this volume demonstrate the diversity of fracture-related articles published by AAPG. Additional AAPG articles are included in a bibliography for the convenience of the reader. Several papers do not deal directly with hydrocarbon reservoirs, but are important to our understanding of fractured reservoirs.

This reprint volume was designed to be a tool for geologists with a limited background in fracturing, yet still be a valuable reference for the geologist experienced in working with fractured reservoirs. The papers provide: (1) definition of fracture terms, (2) theory of fracturing (cause-effect relationships), (3) geologic and reservoir characteristics necessary for fracture-associated production, (4) diagnostic features of a fractured reservoir, (5) relationship of fractures to production, (6) documentation of fracturing, and (7) techniques for the exploration and development of fractured reservoirs.

Because of the significant reserves remaining to be found in fractured reservoirs, it is antici-

pated that this subject will attract new attention and consideration by petroleum geologists.

John R. Kostura
Texaco Inc.
Midland, Texas

June 22, 1977

John H. Ravenscroft
Texaco Inc.
Tulsa, Oklahoma

June 22, 1977

Selected References of Other Fracture Related Articles by the AAPG

Berry, Frederick A. F., 1973, High fluid potentials in California Coast Ranges and their tectonic significance: AAPG Bull., v. 57, p. 1219-1249.

Braunstein, Jules, 1953, Fracture-controlled production in Gilbertown Field, Alabama: AAPG Bull., v. 37, p. 245-249.

Brown, C. W., 1961, Comparison of joints, faults, and airphoto linears: AAPG Bull., v. 45, p. 1888-1892.

Currie, J. B., 1977, Significant geologic processes in development of fracture porosity: AAPG Bull., v. 61, p. 1086-1089.

Daniel, E. J., 1954, Fractured reservoirs of the Middle East: AAPG Bull., v. 38, p. 774-815.

Handin, John, and Rex V. Hager, Jr., 1957, Experimental deformation of sedimentary rocks under confining pressure—tests at room temperature on dry samples: AAPG Bull., v. 41, p. 1-50.

——— ——— M. Friedman, and J. N. Feather, 1963, Experimental deformation of sedimentary rocks under confining pressure—pore pressure tests: AAPG Bull., v. 47, p. 717-755.

Hanna, M. A., 1953, Fracture porosity in Gulf Coast: AAPG Bull., v. 37, p. 266-281.

Harris, J. F., G. L. Taylor, and J. L. Walper, 1960, Relation of deformational fractures in sedimentary rocks to regional and local structures: AAPG Bull., v. 44, p. 1853-1873.

Harris, S. A., 1975, Hydrocarbon accumulation in "Meramec-Osage" (Mississippian) rocks, Sooner Trend, northwest-central Oklahoma: AAPG Bull., v. 59, p. 633-664.

Hunter, C. D., and D. M. Young, 1953, Relationship of natural gas occurrence and production in eastern Kentucky (Big Sandy gas field) to joints and fractures in Devonian bituminous shale: AAPG Bull., v. 37, p. 282-299.

Kerr, P. F., and O. C. Kopp, 1958, Salt-dome breccia: AAPG Bull., v. 42, p. 548-560.

Lattman, L. H., and R. P. Nickelsen, 1958, Photogeologic fracture-trace mapping in Appalachian Plateau: AAPG Bull., v. 42, p. 2238-2245.

—— and A. V. Segovia, 1961, Analysis of fracture trace pattern of Adak and Kagalaska Islands, Alaska: AAPG Bull., v. 45, p. 249-263.

Levorsen, A. I., 1953, Discussion of fractured reservoir subjects: AAPG Bull., v. 37, p. 314-330.

London, W. W., 1972, Dolomite in flexure-fractured petroleum reservoirs in New Mexico and Colorado: AAPG Bull., v. 56, p. 815-821.

Lucas, P. T., and J. M. Drexler, 1976, Altamont-Bluebell—a major, naturally fractured stratigraphic trap, Uinta basin, Utah, in Jules Braunstein, ed., North American oil and gas fields: AAPG Mem. 24, p. 121-135.

McCaleb, J. A., and D. A. Wayhan, 1969, Geologic reservoir analysis, Mississippian Madison Formation, Elk Basin field, Wyoming-Montana: AAPG Bull., v. 53, p. 2094-2113.

McQuillan, Henry, 1973, Small-scale fracture density in Asmari Formation of southwest Iran and its relation to bed thickness and structural setting: AAPG Bull., v. 57, p. 2367-2385.

—— 1974, Fracture patterns on Kuh-e Asmari anticline, southwest Iran: AAPG Bull., v. 58, p. 236-246.

Nelson, R. A., and J. Handin, 1977, Experimental study of fracture permeability in porous rock: AAPG Bull., v. 61, p. 227-236.

Pirson, S. J., 1953, Performance of fractured oil reservoirs: AAPG Bull., v. 37, p. 232-244.

Regan, L. J., Jr., 1953, Fractured shale reservoirs of California: AAPG Bull., v. 37, p. 201-216.

Truex, J. N., 1972, Fractured shale and basement reservoir, Long Beach Unit, California: AAPG Bull., v. 56, p. 1931-1938.

Waldschmidt, W. A., P. E. Fitzgerald, and C. L. Lunsford, 1956, Classification of porosity and fractures in reservoir rocks: AAPG Bull., v. 40, p. 953-974.

Walters, R. F., 1953, Oil production from fractured pre-Cambrian basement rocks in central Kansas: AAPG Bull., v. 37, p. 300-313.

BULLETIN OF THE AMERICAN ASSOCIATION OF PETROLEUM GEOLOGISTS
VOL. 33, NO. 1 (JANUARY, 1949), PP. 32-51, 15 FIGS.

FRACTURED RESERVOIRS OF SANTA MARIA
DISTRICT, CALIFORNIA[1]

LOUIS J. REGAN, JR.,[2] AND ADEN W. HUGHES[3]
Bakersfield and Santa Maria, California

ABSTRACT

Cumulative production of the Santa Maria district to January 1, 1947, was 250,637,000 barrels, of which an estimated 77 per cent was from fractured rocks and 23 per cent from oil sands. Seventy-four and one-half per cent of the oil was produced from fractured rocks in the Monterey formation and 2 per cent from fractured Knoxville sandstones. Monterey fractured rocks in order of importance include (1) Mohnian cherts, (2) Luisian calcareous shale, and (3) Mohnian platy siliceous and porcellaneous shale. The distribution, age, and character of fractured zones are illustrated by stratigraphic sections. The distribution of maximum chert development suggests chert originated as a sedimentary facies. General characteristics of potentially productive fractured rocks of the district are analyzed and their permeabilities and porosities, as indicated by production data, are discussed.

INTRODUCTION

The Santa Maria district produced 250,637,000 barrels of oil to January 1, 1947. In all, 192,000,000 barrels, or 77 per cent was produced from various types of fractured rock and only 23 per cent was produced from sand. The district offers a classic example of production from fractured rocks and a fertile field for the study of types and characteristics of rocks which are potentially fractured.

Figure 1 illustrates the stratigraphic position of the types of reservoirs in the district and the economic importance of each.

The stratigraphic position of oil sand zones is indicated by Column 1.

Column 2 portrays the position of fractured shale reservoirs, a group of zones which has produced 74.2 per cent of the oil. Fractured shale zones are productive in all fields of the district.

Column 3 illustrates the stratigraphic position of fractured sandstone zones. These include (1) fractured calcareous sandstones of the Point Sal formation, productive to a very minor degree at the east end of the Santa Maria Valley field, (2) fractured calcareous sandstones of the Lospe, productive in the Arrellanes pool of the Casmalia field, and (3) most important, fractured calcareous sandstones referred to the Knoxville[4] formation and productive in the north-central part of the Santa Maria Valley field. The last two sandstones are older than known local source beds. The accumulation of oil in these rocks is similar to the fractured basement accumulations in other districts. These zones have produced 2.4 per cent of the oil in the Santa Maria district.

Some of the characteristics of fractured zones in the district are as follows.

---

[1] Read before the Pacific Section at Pasadena, California, November 6, 1947. Manuscript received, August 26, 1948. Published by permission of C. M. Wagner, General Petroleum Corporation, and L. N. Waterfall, Union Oil Company of California. Credit is due A. J. Taylor, Santa Maria, California, for the photography.

[2] Geologist, State Exploration Company, Bakersfield.

[3] Paleontologist, Union Oil Company of California, Santa Maria.

[4] In this article, Knoxville is used tentatively in correlating these sandstones.

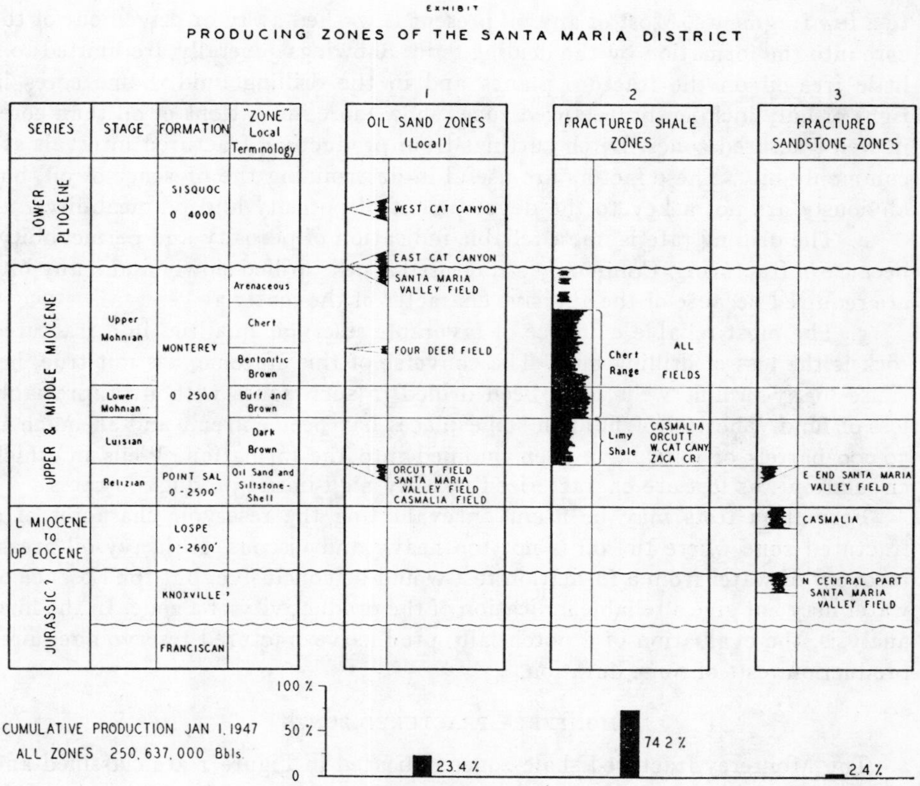

EXHIBIT

PRODUCING ZONES OF THE SANTA MARIA DISTRICT

SERIES	STAGE	FORMATION	"ZONE" Local Terminology	1 OIL SAND ZONES (Local)	2 FRACTURED SHALE ZONES	3 FRACTURED SANDSTONE ZONES
LOWER PLIOCENE		SISQUOC 0 4000		WEST CAT CANYON		
				EAST CAT CANYON		
UPPER & MIDDLE MIOCENE	Upper Mohnian	MONTEREY 0 2500	Arenaceous	SANTA MARIA VALLEY FIELD		
			Chert			
			Bentonitic Brown	FOUR DEER FIELD	Chert Range ALL FIELDS	
	Lower Mohnian		Buff and Brown			
	Luisian		Dark Brown		Limy CASMALIA Shale ORCUTT W CAT CANY ZACA CR.	
	Relizian	POINT SAL 0 - 2500'	Oil Sand and Siltstone - Shell	ORCUTT FIELD SANTA MARIA VALLEY FIELD CASMALIA FIELD		E END SANTA MARIA VALLEY FIELD
L MIOCENE TO UP EOCENE		LOSPE 0 - 2600'				CASMALIA
JURASSIC ?		KNOXVILLE ?				N CENTRAL PART SANTA MARIA VALLEY FIELD
		FRANCISCAN				

CUMULATIVE PRODUCTION JAN 1. 1947
ALL ZONES 250. 637. 000 Bbls.

100 % — 50 % — 0

23.4 %   74.2 %   2.4 %

FIG. 1.—Producing zones of Santa Maria district.

1. Potentially fractured reservoirs, meaning zones most susceptible to fracturing, are composed of very brittle, hard, generally conchoidally fracturing rocks. These rocks include in order of importance: (1) chert, (2) calcareous shales and shells, (3) platy siliceous shale, and (4) sandstones. It appears that the harder and more brittle rocks have better potential development of secondary porosity and permeability by fracturing.

2. Productive fractured rocks commonly exhibit favorable reservoir character on the self-potential curve of an electric log. The amount of relief shown by the curve can not be taken as an index of productivity of a well, but a very flat self-potential curve indicates either a tight, poorly fractured zone, or one in which the fractures are sealed by tar. The resistivity of a fractured zone in this district is generally high on all curves regardless of fluid content because of the dense character of the rock.

3. Indications of oil when drilling and coring potentially productive fractured reservoirs are unimpressive. Core recovery is commonly poor and may be limited

to a few fragments. Most of any oil present is washed away or driven out of the core into the formation by the drilling fluid. Showings generally are limited to a little free oil on the fracture planes and in the drilling fluid. Some cores in tight or only incipiently fractured zones show more indications of oil than cores in well fractured zones. Ditch cuttings from productive, fractured intervals are commonly oily. These factors are useful in determining the presence of oil, but obviously are not a key to the development of porosity and permeability.

4. The drilling rate is not a reliable indication of porosity and permeability, because of fracturing. Commonly a fractured zone is drilled slowly and many bits are required because of the abrasive character of the material.

5. The most reliable evidence of favorable reservoir qualities in a fractured rock is the loss of drilling fluid. The converse of this statement is not true, because many prolific wells have been drilled in such zones without appreciable loss of fluid. The loss of fluid in some places has been extreme and as much as 30,000 barrels of fluid have been pumped into the formation. Wells in which circulation was lost are characterized by very high initial production rates.

Formation tests may be useful in evaluating the reservoir character of a fractured zone where the oil is not too heavy and viscous. In heavy-oil areas, recovery of water from a formation test would be conclusive, but the absence of water may not give a reliable indication of the productivity of a zone. In the final analysis, the evaluation of a potentially productive fractured reservoir requires production tests of some duration.

MONTEREY FRACTURED ZONES

The Monterey fractured shale zones indicated in Figure 1 are classified and described in order of economic importance: (1) chert zones, (2) calcareous shale zones, and (3) platy siliceous and porcellaneous shale zones.

1. The chert zones are confined to the Mohnian stage. The character, development, and age of the chert in this stage are variable, and a generalized description of the zones can best be accomplished by the use of three type sections: (A) the Santa Maria Valley field, (B) the West Cat Canyon field, and (C) the Lompoc field.

A. The Santa Maria Valley chert is upper Mohnian in age and occupies the "cherty zone" and part of the "bentonitic brown zone" of local terminology. It consists of interbedded brown opaline chert and brown and black siliceous shale with rare streaks of light gray to white opaline chert in the lower part. The actual percentage of chert has not been determined because of poor core recovery, but chert is estimated to comprise an average of less that 50 per cent and locally as low as 10 per cent of the section. The appearance of an unfractured part of a typical core in the Santa Maria Valley chert is illustrated by Figure 2 which shows the dark brown banded appearance of the Santa Maria Valley type chert. Figure 3 shows the fractured light gray to white opaline chert and siliceous shale.

B. The West Cat Canyon chert is lower Mohnian in age. This section shows greater values on electric logs and larger amounts of chert in ditch samples than the other type sections. Figure 4 illustrates part of a core from the fractured chert zone at West Cat Canyon. The West Cat Canyon chert section consists of interbedded dark brown to black siliceous shale, gray white and black opaline chert, buff phosphatic shale, and lentils of very massive buff siliceous limestone. Opaline chert appears to comprise 50–60 per cent of the lower Mohnian section at West Cat Canyon, based on counts of chert fragments in ditch samples. The amount of siliceous limestone in the lower Mohnian has not been determined. It is gray-brown to buff in color, massive, very hard, and in cores is generally moderately fractured. Oil is contained in small solution cavities in rocks of this type, as indicated by Figure 5, and secondary permeability of this type may be locally important.

C. In the Lompoc field, chert is in both the upper and lower Mohnian, and a complete section of Monterey is present. A thin zone of brown shale is present at the top and is underlain by three fairly distinct cherty members. The lowest chert member is similar to the Cat Canyon chert with which it is correlated. The upper two cherty members are believed to be upper Mohnian in age and equivalent to the Santa Maria Valley chert, but are lithologically dissimilar to other sections. For the purpose of this paper, the two upper cherty members are called the Lompoc type cherty shale.

The Lompoc type cherty shale consists of 90–95 per cent thin-bedded, brown siliceous shale, with rare thin lenses of white and brown opaline chert and buff siliceous limestone. The lower part consists of gray to light gray siliceous to porcellaneous shale interbedded with light gray and white opaline chert, with chert more abundant.

2. The calcareous shale type of fractured zone is in Luisian rocks in much of the district, and consists essentially of brown silty phosphatic shale with interbeds of calcareous shale and limestone. The calcareous streaks give large values to the resistivity curve and relatively small throws on the self-potential curve of electric logs. This interval is in places identified as chert on the basis of electric logs, but ditch samples and cores indicate little or no chert in the zone. The relative importance of these rocks is not known because of the absence of selective tests, but cores in the zone at favorable structural position show oil. The interval is open in many structurally high field wells.

3. Platy siliceous and porcellaneous shales are of minor importance as fractured reservoirs. These shaly members of the upper Mohnian are sufficiently fractured to be productive in the Santa Maria Valley, Casmalia, Orcutt, and West Cat Canyon fields. The permeability in this type of material is generally not sufficient for commercial production of the typical heavy Santa Maria crude. Lighter oil of the Orcutt area has been successfully produced from this type of fractured zone in the newly discovered Four Deer area. The distribution, variations, and limits of types of Monterey fractured zones are illustrated by stratigraphic section (Figs. 7–11). The locations of the sections are shown on the index map (Fig. 6),

which also indicates the position of the Santa Maria oil fields and structure contours on the base of the Pliocene in the fields. The Monterey basin is delineated on this map. The areas considered as positive or nearly positive during Monterey time are the following.

  1. Nipomo Mesa north and northeast of Santa Maria Valley field
  2. Basement ridge in the vicinity of East Cat Canyon field
  3. La Laguna basement high east of Gato Ridge field
  4. Santa Ynez uplift which marks south limit of Santa Maria basin
  5. Point Sal uplift

Stratigraphic section AA' (Fig. 7) is an approximately west-east section from the Point Sal positive area through the center of the district to the basement ridge in the East Cat Canyon area, a distance of 17 miles. Monterey type sections of the Casmalia, Orcutt, and Cat Canyon fields are shown in section AA'. The character and age of Monterey fractured zones are fairly uniform east and west through the center of the district. This is well indicated by an electric-log correlation, although the distance between wells is great, and also by the position of the top of the lower Mohnian. The chert in this part of the district is mainly the lower Mohnian Cat Canyon type or type 1-B. This chert is thickly developed in the West Cat Canyon field, thins across the Orcutt structure, and reaches maximum development in the Casmalia field. The chert thins northwest of the Casmalia field, approaching the Point Sal positive area. Thin stringers of the upper Mohnian Santa Maria Valley type chert, type 1-A, appear in sections at Orcutt and West Cat Canyon.

The Luisian calcareous shale fractured zone, type 2 of the present classification, is well developed in the area of this section. The electric-log character of this zone is similar to the chert zone, with somewhat more subdued throws on the self-potential curve. Few isolated tests have been made of this zone.

Platy siliceous shale 200 feet above the chert zone, and described under type 3, has a somewhat permeable electric-log characteristic. This zone is productive in the Orcutt field, and reportedly produces a small amount of heavy oil in the Casmalia field.

The relatively simple stratigraphy and rather uniform distribution of fractured zones shown by the west-east section are in contrast to a more complex pattern exhibited by north-south diagrammatic sections.

Stratigraphic section BB' (Fig. 8) is the most westerly of the north-south sections, and extends from the west end of the Santa Maria field through the Casmalia field to the Lompoc field, a distance of 14 miles. The most northerly log, No. 1 on the section, is representative of a well developed chert section of the Santa Maria Valley type, or type 1-A. The chert is progressively overlapped toward the north by Pliocene Sisquoc. The second log on the south, No. 2, shows the same type of chert in the upper Mohnian 2.5 miles south of the Santa Maria Valley field. The lower Mohnian has not been penetrated in this area. The upper Mohnian type 1-A chert is replaced in the Casmalia field by moderately fractured,

FIG. 3.—Type 1-A, Santa Maria Valley type chert (fractured).

FIG. 2.—Type 1-A, Santa Maria Valley type chert (unfractured).

FIG. 4.—Type 1-B, Cat Canyon type chert.

FIG. 5.—Type 1-B, siliceous limestone in Cat Canyon type chert.

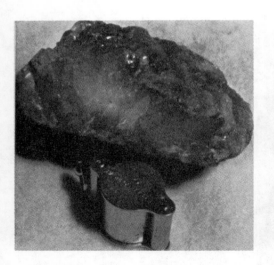

FIG. 12.—Secondary quartz recovered by bailer from Cat Canyon well.

FIG 13.—Outcrop sample of type 1-B lower Mohnian chert.

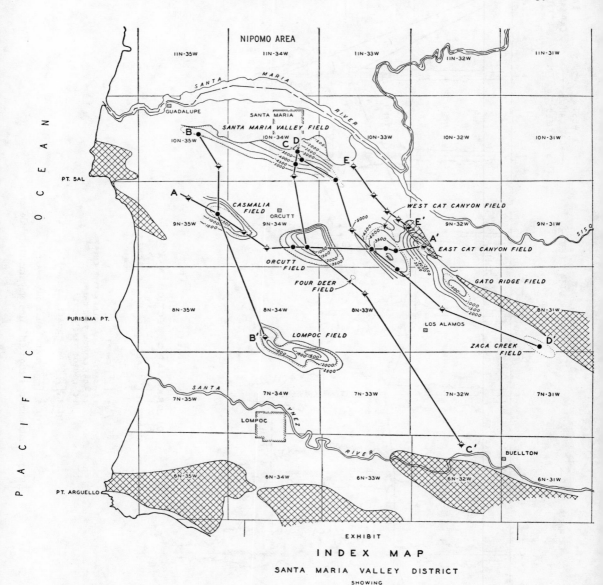

Fig. 6.—Index map, Santa Maria Valley district.

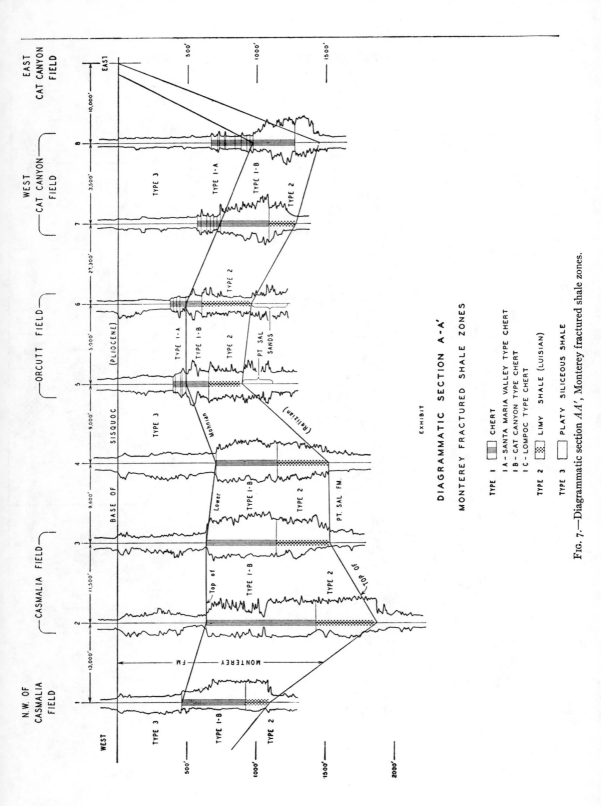

FIG. 7.—Diagrammatic section *AA'*, Monterey fractured shale zones.

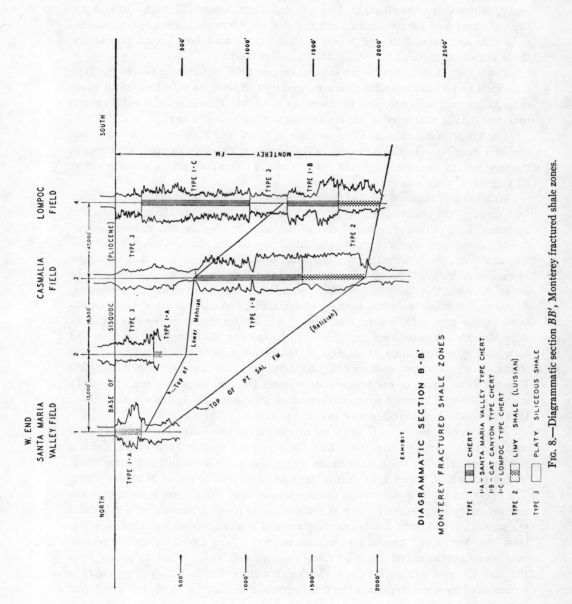

Fig. 8.—Diagrammatic section *BB'*, Montérey fractured shale zones.

platy, thin-bedded brown shale. The well developed lower Mohnian Cat Canyon type of chert and Luisian calcareous shale of the Casmalia field are represented by a thin section of buff and brown siliceous shale and dark brown phosphatic shale at the west end of the Santa Maria Valley field.

The upper Mohnian is cherty in the Lompoc field, but this section is the type described as 1-C above and is characterized by thin-bedded interbedded siliceous shale, siliceous limestone, and thin streaks of chert. This lithologic difference is reflected by the character of the electric logs in the Lompoc field.

The lower Mohnian Cat Canyon type of chert and Luisian calcareous shale are well developed in the Lompoc field. However, the only productive zones in the field are in the upper Mohnian chert, and the lower fractured zones are wet at all structural positions.

Stratigraphic section *CC'* (Figure 9) extends from the Santa Maria Valley field through the Orcutt field, to the south side of the Purisima Hills, a distance of 23 miles. Two wells at the north end of the sections, Nos. 1 and 2, encountered thickly developed chert of the upper Mohnian Santa Maria Valley field type. The next well, No. 3, a southern outpost test of the field, illustrates that the upper Mohnian chert grades into siliceous shale, and has the first appearance of the lower Mohnian Cat Canyon type of chert. Luisian calcareous shale is also present. The Orcutt section is similar to the last described, but chert is very poorly developed in the upper Mohnian. Well No. 5, 3 miles southeast of the Orcutt field, encountered a similar section, but with a thicker development of the Cat Canyon type chert. The Monterey on the south side of the Purisima Hills is similar to the Lompoc section (Fig. 8), but differs in that the upper Mohnian Lompoc type chert is very poorly developed and has been replaced by platy siliceous shale. The most interesting feature of the section is the clear evidence that the thick chert is younger from south to north.

Stratigraphic section *DD'* (Fig. 10) extends from the Santa Maria Valley field southeasterly through the West Cat Canyon field to the Zaca Creek field, a distance of 21 miles. On the north, two representative wells in the Santa Maria Valley field, Nos. 1 and 2, exhibit a thick development of upper Mohnian chert. The chert grades into platy shale southeast, and very little chert is present in the southeasterly outpost well No. 3. A few streaks of lower Mohnian type 1-B chert are present in this well. A large development of the lower Mohnian chert extends from the West Cat Canyon field southeast to the Zaca Creek field. The section in the well northwest of the West Cat Canyon field, No. 4, is not as cherty as in the Cat Canyon field. The upper Mohnian Santa Maria Valley type chert is represented by a few streaks in the Cat Canyon area, but is also thickly developed in the Zaca Creek field.

Stratigraphic section *EE'* (Fig. 11) extends from a well near the east end of the Nipomo positive area, southeast to the East Cat Canyon positive area, a distance of 6 miles. Well No. 1, near the east end of the Nipomo high, encountered

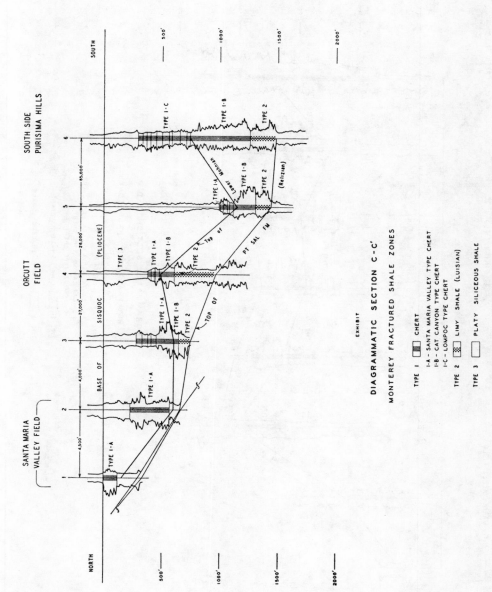

DIAGRAMMATIC SECTION C-C'

MONTEREY FRACTURED SHALE ZONES

TYPE 1 ▨ CHERT

1-A — SANTA MARIA VALLEY TYPE CHERT
1-B — CAT CANYON TYPE CHERT
1-C — LOMPOC TYPE CHERT

TYPE 2 ▨ LIMY SHALE (LUISIAN)

TYPE 3 ☐ PLATY SILICEOUS SHALE

Fig. 9.—Diagrammatic section CC', Monterey fractured shale zones.

LOUIS J. REGAN, JR., AND ADEN W. HUGHES

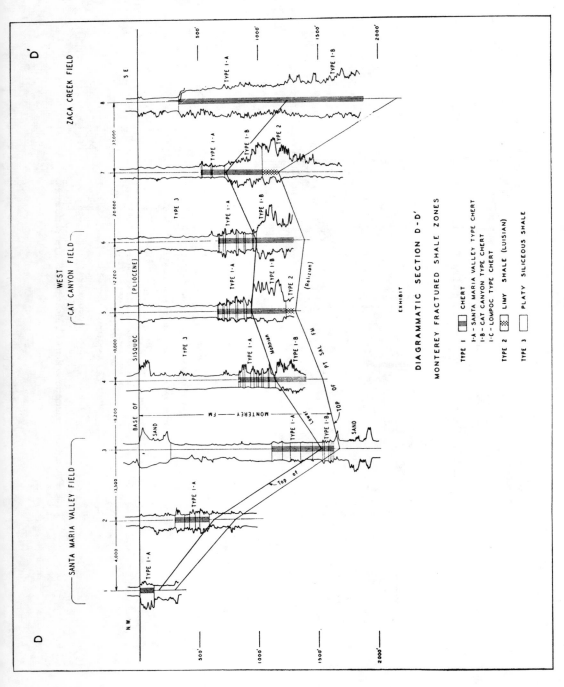

FIG. 10.—Diagrammatic section *DD'*, Monterey fractured shale zones.

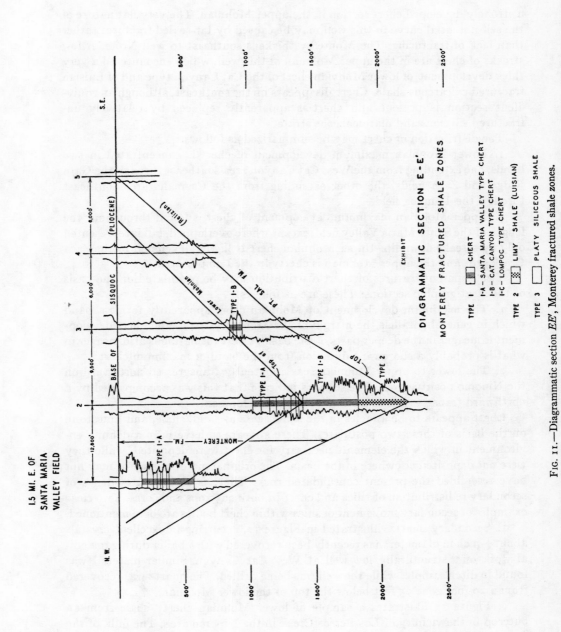

FIG. 11.—Diagrammatic section *EE'*, Monterey fractured shale zones.

a strongly developed chert section in the upper Mohnian. The very flat nature of the self-potential curve in this well may be caused by tar-sealed fractures rather than lack of fracturing. The Monterey thickens southeast to well No. 2. A few streaks of chert are in the upper Mohnian of this well, which encountered a very thick development of lower Mohnian chert of the Cat Canyon type and of Luisian fractured calcareous shale. Chert disappears on the southeast, although an equivalent section is present and chert is apparently replaced by relatively unfractured siliceous and diatomaceous strata.

The distribution of chert may be summarized as follows.

1. Lower Mohnian maximum development of chert is concentrated in two bands, one extending from the West Cat Canyon area southeast through the Gato Ridge and Zaca fields, the other extending from the Casmalia area southeast through the Lompoc field.

2. Upper Mohnian maximum development of chert is found throughout the length of the Santa Maria Valley field, except where overlapped, but is represented only by streaks downdip. Upper Mohnian chert is highly developed in the Zaca Creek area and the upper Mohnian is cherty in the Lompoc area.

Two apparent features of chert distribution may be postulated on the basis of the stratigraphic sections. These are as follows.

1. The maximum development of Mohnian chert apparently forms a band which in general parallels the margin of the Monterey basin, and chert development is meager in the deeper parts of the basin. Chert formation is also meager in what is probably a shoreward direction from the band of maximum chert.

2. The lower to upper Mohnian seas apparently transgressed northward on the Nipomo positive area. The sections indicate that this was accompanied by a northward transgression of maximum chert development.

Chert appears to be a facies in the same sense as sand, shale, and limestone on the basis of these two postulates. There appears to have been a definite environment in which the elements necessary for chert were deposited, while they were not deposited elsewhere in the basin. The original material may or may not have resembled the present consolidated rock. There is abundant evidence of secondary redistribution of silica and chert in both outcrops and cores. Two clear examples of secondary movement of silica within chert bodies are here mentioned.

1. Secondary quartz, illustrated in Figure 12, occurring in excellent crystals about $\frac{1}{4}$-inch in diameter, has recently been recovered with a bailer during remedial work on a structurally low well at West Cat Canyon. Similar material was found in ditch samples while the well was being drilled. The quartz was recovered from a 20-foot zone 65 feet below the top of the lower Mohnian.

2. Figure 13 illustrates a sample of lower Mohnian cherty shale from an outcrop in the vicinity of Los Berros Creek in the Nipomo area. The bulk of the sample is composed of white opaline chert and porcellaneous shale, which is cut by lenses and veinlets of brown opaline chert.

There are no clear examples in the Santa Maria district where chert is well developed and is not sufficiently fractured to possess some porosity and permeability. Some fractures may have been sealed by tar where cherty sections have not yielded fluid on tests. Fractured zones are probably the result of adjustments in brittle rocks to tectonic movements. Theoretically, highly fractured areas should be related to structural features, such as anticlinal axes or faults. No such relationship can be demonstrated in the Santa Maria district at present. It appears that rocks sufficiently brittle to be susceptible to fracturing are regionally fractured and broken and are potential reservoirs. Variations in fracturing and in consequent size of wells are extreme and unpredictable locally in Santa Maria oil fields. A property in the southeast part of the Santa Maria Valley field is an example of this. The property is drilled with one well to 10 acres. Initial production of these wells varied from the maximum of 2,500 barrels per day to the minimum of 90 barrels per day. Cumulative production of the wells varies between the minimum of 12,000 barrels to the maximum of 500,000 barrels. A direct offset well to the property is reported to have produced more than a million barrels and is still believed capable of producing 1,200 barrels per day. These figures may be somewhat misleading because wells were not completed at the same time, but they are believed to reflect a great variation in fracturing in the chert zone.

Initial production rates of 2,500 barrels per day were common, and potentials as high as 10,000 barrels per day were established in the early history of the Santa Maria Valley field. These high rates of production of heavy viscous oil indicate high permeability. An estimated maximum permeability of 35 darcys and an average permeability of 10–15 darcys have been obtained empirically for the chert zone of the Santa Maria Valley field by comparison with oil sand zones in various California oil fields. The estimates are based on potentials of oil sands of known permeability in zones where thickness and pressure are similar to the Santa Maria Valley field. Santa Maria crude is much more viscous than the oil produced from sands used for comparison. The permeability of sands used for comparison is a maximum of 2–3 darcys. The much higher estimated permeability of the chert zone is the result of adjusting the permeability of comparable sand zones to fit the viscosity of Santa Maria Valley field crude at bottom-hole temperatures. Similar but somewhat smaller values for the permeability of fractured chert have been obtained from productivity indices by the method described by Norris Johnston.[5] The permeability of sands and fractured reservoirs are not strictly comparable because of physical characters such as surface tension and capillary action, but it seems clear that the permeability of fractured chert is generally high and in some places extreme.

Two outstanding effects of the high permeability are the following.

1. The production of a well is commonly influenced by an offset well. In some places, fluid lost in a well is produced by offset wells. In one well a casing failure

[5] Norris Johnston and John Sherborne, "Permeability as Related to Productivity Index," *A.P.I. Drilling and Production Practice* (1943).

developed, giving fresh water access to the chert zone. Water cuts in an adjacent well reflected the condition immediately.

2. A common development of a very unequal encroachment of edge water. Two outstanding examples of this condition are the present irregular oil-water interfaces along the south-central part of the Santa Maria field, and at the northwest end of the West Cat Canyon field.

A well developed fractured chert zone may ultimately produce 20,000–40,000 barrels per acre or approximately 100–150 barrels per acre foot. It has been estimated that producible oil in the chert zone may be as high as 30 per cent of the total oil in the formation. The average porosity of such a zone is 6 per cent on the basis of these figures.

Fractured rocks form satisfactory reservoirs for oil of heavy to very heavy gravity because of the high permeabilities and because sand problems characteristic of low-gravity oil sands are not encountered. It seems probable that lighter crudes could be produced commercially from zones exhibiting less fracturing.

KNOXVILLE PRODUCTION

The basement production of the Santa Maria Valley field was discovered in December, 1942, by W. R. Gerard's Acquistapace well No. 1 which initially flowed 1,200 barrels per day of 18.5° gravity oil. This well was in an area considered sub-commercial in Monterey zones. It encountered 250 feet of Monterey, with 120 feet of cherty shale below the Sisquoc. Approximately 30 feet of detrital or weathered material was found at the top of the Knoxville, and 90 feet of Knoxville fractured sandstones were penetrated. The oil in the Knoxville is identical with Monterey oil and is believed to have migrated from the Monterey into fractured members of the older formation.

Both the Monterey and older rocks are open to production in the discovery well and in most wells penetrating the Knoxville. Therefore, it is difficult to evaluate the extent of basement production because of the lack of selective tests. Wells which produce from the Knoxville include (1) wells drilled north of the Monterey limits and productive from Knoxville alone, (2) wells deepened into the Knoxville with an increased productivity, and (3) wells completed in both Monterey and Knoxville with production greatly in excess of rates expected from Monterey alone. Wells which prove the Knoxville non-productive include (1) wells north of Monterey limits which lack fluid in the Knoxville, (2) wells lacking fluid in both Knoxville and Monterey, (3) wells deepened into the Knoxville without change in productivity, and (4) wells penetrating both Knoxville and Monterey with production rate similar to Monterey zones.

Figure 14 is a generalized structure map of the Santa Maria Valley field, contoured on top of the Knoxville and showing locations of productive and non-productive wells. Productive Knoxville wells are limited to the north-central part of the field. The west end of the field is not favorable for basement production because the oil-water interface common to both formations occurs within the

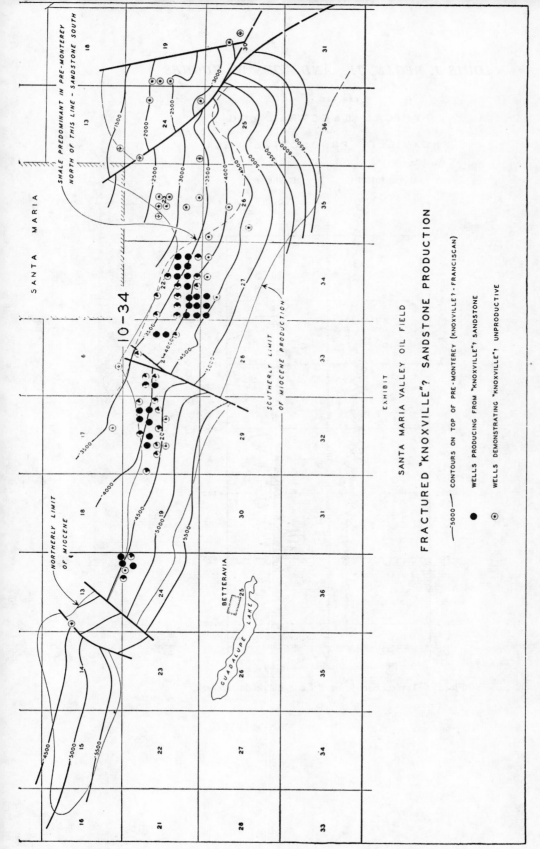

SANTA MARIA VALLEY OIL FIELD

FRACTURED "KNOXVILLE"? SANDSTONE PRODUCTION

EXHIBIT

—5000— CONTOURS ON TOP OF PRE-MONTEREY (KNOXVILLE?-FRANCISCAN)

● WELLS PRODUCING FROM "KNOXVILLE"? SANDSTONE

⊛ WELLS DEMONSTRATING "KNOXVILLE"? UNPRODUCTIVE

FIG. 14.—Santa Maria Valley oil field, fractured Knoxville sandstone production.

21

EXHIBIT

TYPICAL ELECTRIC LOG
OF
"KNOXVILLE" PRODUCTIVE WELL

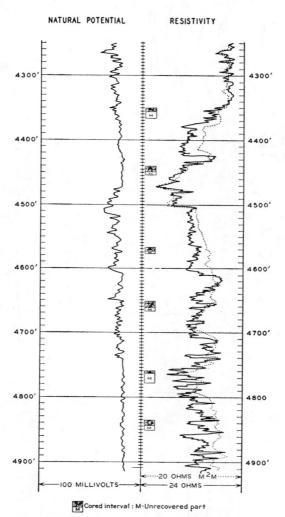

Fɪɢ. 15.—Typical electric log of Knoxville productive well.

Monterey. Some wells have found the Knoxville non-productive at the east end of the field. These wells encountered mainly gray and black, micromicaceous, slickensided, plastic-appearing shale, apparently without reservoir characteristics, while productive wells encountered dominantly sandstone. Casts of the megafossil *Aucella* have been reported in the shales at the east end of the field, and on the basis of these the zone is referred to as the Knoxville. No fossils have been reported in the productive area. The approximate northern limit of fractured Knoxville sandstone is indicated on the map. The northern limit of production is believed to be controlled partly by overlap of the fractured sandstone by the Sisquoc, and partly by tar seal, at the contact.

Figure 15 is a typical electric log of a well producing from the Knoxville, showing the section penetrated below the Miocene. The well was originally completed in the Monterey in August, 1942, initially producing 25 barrels per day. In its original condition, the total depth was 4,332 feet after penetrating 90 feet into the Knoxville. The liner was landed at 4,234 feet or at the top of the basement without a bottom plug. The last production before deepening was 13 barrels per day. The well was deepened to 4,910 feet, equivalent to 665 feet of penetration into the Knoxville, and was recompleted December 3, 1945, pumping 220 barrels per day of 17.1° gravity clean oil. Present production of the well is 90 barrels per day, and it has produced a cumulative production of approximately 70,000 barrels of oil from the Knoxville. Cores and ditch samples indicate the well encountered dominantly sandstone. The sandstone is light blue-gray, medium- to fine-grained, micaceous, pyritiferous, very calcareous, with veinlets of calcite. Fractures were present in most of the cores, spaced from a few inches to 3 feet apart, with a coating of oil on the fracture surfaces. The sandstone was generally oil-stained in a width of $\frac{1}{4}$–$\frac{1}{2}$ inch from the fractures. Live oil was present in cores throughout the interval, but the gravity of the oil appeared to increase with deeper penetration.

SUMMARY

It appears that the brittle hard rocks of the Santa Maria district are regionally fractured, but show extreme local variation in amount of fracturing. Chert is the most important fractured rock, but calcareous shale, platy siliceous shale, and sandstone are sufficiently fractured to form commercial oil reservoirs. The main zones susceptible to fracturing have wide distribution in the district. The distribution of maximum chert development suggests that the chert originated as a sedimentary facies. Fractured chert reservoirs are characterized by low porosity and high permeability. Expected recovery per acre foot in such zones is relatively low.

A knowledge of the characteristics of fractured rock reservoirs, observed in the course of drilling, is an aid to the evaluation of the productive possibilities of such a reservoir; but in the final analysis production tests are required.

Copyright 1953 by The American Association
of Petroleum Geologists
BULLETIN OF THE AMERICAN ASSOCIATION OF PETROLEUM GEOLOGISTS
VOL. 37, NO. 2 (FEBRUARY, 1953), PP. 217-231, 10 FIGS.

DILATANCY IN MIGRATION AND ACCUMULATION
OF OIL IN METAMORPHIC ROCKS[1]

DUNCAN A. McNAUGHTON[2]
Houston, Texas

ABSTRACT

The difficulties inherent in the prediction of potential reservoir conditions in basement rocks buried by a concealing mantle of sediments are apparent to all geologists and constitute the main cause for prejudice against basement exploration. Consequently, geologists have not played an important role in the discovery of oil accumulations in basement rocks. All the major discoveries have been accidental.

Our record can be improved by a critical examination of our working hypotheses concerning the nature of the geological processes responsible for the formation of basement oil pools. Among these hypotheses is the generally accepted "up-slope" theory of oil migration in basement reservoirs. The theory is closely patterned after our working concept of migration of oil in conventional sandstone carrier beds. The theory and its limitations as a guide for exploratory work are discussed.

An alternative proposition is advanced that fracturing of competent basement rocks involves dilatancy which in turn reduces hydrostatic pressures in focal areas of deformation. Pressure gradients are thereby established between the potential basement reservoir rocks and the overlying source and carrier beds containing oil, gas, and water. Thus a tendency to "suck in" fluids into the basement rocks is established.

Deposition of calcite in fractured facies of the basement complex above some oil accumulations may be due indirectly to dilatancy—reduced pressures allowing carbon dioxide to escape from waters associated with oil and gas thus causing precipitation of calcium carbonate.

INTRODUCTION

This paper is concerned primarily with the geological processes responsible for the migration and accumulation of oil in the basement rocks underlying five oil fields in southern California.

The objective of the paper is to arouse interest in these unusual accumulations of oil and thus to foster discussion which may lead eventually to better understanding of the processes involved in their formation.

HISTORY OF BASEMENT OIL DISCOVERIES IN CALIFORNIA

The difficulties inherent to the prediction of potential reservoir conditions in basement rocks buried by a concealing mantle of sediments are apparent to all geologists and constitute the main cause for prejudice against this type of exploration. Prior to the discovery of the large basement pool in Edison field in 1945, this prejudice was augmented by the conviction that oil in basement rocks

[1] Read before the Association at Los Angeles, March 26, 1952. Manuscript received, April 30, 1952. Published by permission of the Gulf Oil Corporation, Pittsburgh, Pennsylvania. Part of a dissertation submitted in partial fulfillment of the requirements for the degree of Doctor of Philosophy at the University of Southern California, June, 1950.

[2] Gulf Oil Corporation. The writer is indebted especially to Glen W. Ledingham who outlined problems associated with basement oil pools and later enlisted the financial support of the Western Gulf Oil Company in the investigation. R. L. Hewitt and John May allowed the writer to study their collection of thin sections obtained from basement rocks in the San Joaquin Valley; J. H. Beach and A. H. Huey contributed valuable data concerning oil production and geological relationships in the Edison oil field; Aden Hughes contributed most of the unpublished information concerning the Santa Maria Valley oil field; and M. Mayuga and R. Winterburn allowed the writer to have free access to unpublished information concerning the Wilmington oil field. Special thanks are due to Thomas Clements, W. H. Easton, and E. H. Powers for helpful criticism of this paper.

217

25

is a vagary of nature possessing some scientific interest but slight economic importance. The dangers of basing exploration on negative theories of this nature are well illustrated in the following review of the history of basement discoveries in California.

Probably the first commercial oil produced from metamorphic rocks in California was in Placerita Canyon, near Newhall, California. According to Brown and

Fig. 1.—Map of southern California showing locations of fields producing oil from fractured metamorphic rocks.

Kew (1932, p. 777), small quantities of light-gravity oil were produced from five wells drilled between 1899 and 1901. The basement complex in the vicinity of the wells consists chiefly of schists.

The next discovery of oil in the basement complex was made in the Playa del Rey field at Venice, California, in 1929. This discovery appears to have been forgotten in the passage of years, for it is not mentioned in recent reports (Huey, 1947, and Cabeen and Sullwold, 1946) dealing with basement discoveries. The following quotation from Barton (1931, p. 11) is of historical interest.

> The schist is the lowermost member of the lower zone. Originally it was not considered as a part of the zone, but since wells deepened into it have increased their production it has had to be taken into consideration.

An important basement discovery was made in El Segundo field in 1937. It was accidental in that the objective of the test well was the schist conglomerate capping the basement. Examination of cores revealed that the well stopped in fractured schist containing oil. The well was highly productive and the discovery was followed by intensive development of the basement reservoir in this field. The significance of the discovery was not appreciated by geologists, according to Cabeen and Sullwold (1946, pp. 17–18).

In spite of the fact that schist wells came in for initial productions of up to 4,500 barrels per day, fractured basement rocks were not accepted as testworthy reservoirs except in the immediate vicinity of El Segundo, where several outlying schist tests were drilled.

Another basement pool was discovered in 1942 in the Santa Maria Valley field. This discovery was also accidental, being the fortunate consequence of ignorance concerning the exact depth of the basement in the central part of the field.

History was repeated in 1945 when similar exploration in the Edison field, southeast of Bakersfield, disclosed oil in fractured metamorphic rocks. Intensive development of the basement reservoir followed this discovery. Beach and Campbell (1946) furnished some interesting statistics covering the results of the first year's work. One hundred and three wells were drilled into the basement and all but six of these were completed as commercial producers. Oil production from these wells during the first 18 months after discovery amounted to 4,500,000 barrels and in 1947 the estimated reserve in the basement pool was 25,000,000 barrels (Huey, 1947).

Another basement oil pool was discovered accidentally in the Wilmington field in 1945.

The results of intensive exploration following the Edison and Wilmington discoveries were discouraging as sustained commercial production of oil from basement rocks was not developed elsewhere in California. Thus, none of the important discoveries can be credited to geology.

Geological prejudice was undoubtedly responsible for our failure to guide exploration work prior to 1945. Since then our failure to find new oil pools of this type is probably due to (1) the paucity of sizable oil accumulations in basement rocks elsewhere in southern California and (2) our inability to recognize areas favorable for the accumulation of oil in these rocks. The latter, in turn, may be due to an incomplete understanding of the processes responsible for the formation of these unusual oil accumulations.

MIGRATION AND ACCUMULATION OF OIL IN BASEMENT ROCKS

The prevailing concept concerning migration and accumulation of oil in basement reservoirs is well illustrated in Figure 2 (Cabeen and Sullwold, 1946, p. 25).

This concept embodies several assumptions including the following.

1. Oil originates in sediments capping the basement. All geologists familiar with the basement pools in southern California will agree that the assumption is well supported by field evidence.

2. Oil migrates up the dip of sediments buttressing against the basement, crosses the unconformity, and continues to move up the slope of the basement to the crest of the buried hill or to some position on the flanks of the structure where its upward movement is stopped by a permeability barrier. Oil accumulations at or near the crest of buried hills underlying the Wilmington, El Segundo, and Playa del Rey fields appear to support this assumption whereas flank accumulations in the Edison and Santa Maria Valley fields must be attributed to hypothetical permeability barriers that are difficult to reconcile with other field evidence. For example, in the Santa Maria Valley field, Regan and Hughes (1949, p. 49) have shown that the basement accumulation lies directly below the basal Miocene producing zone (Fig. 3) and is not at higher structural positions where

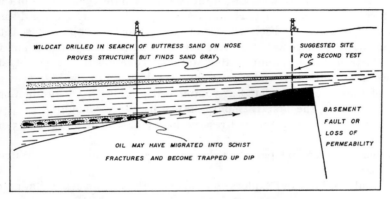

FIG. 2.—Hypothetical example of possible schist accumulation in "condemned" area. After Cabeen and Sullwold (1946, p. 25).

the non-productive Sisquoc formation of Pliocene age rests on the basement. Similarly in the Edison field, factors such as brecciation adjacent to the Edison fault system (Fig. 8), the nature of sediments capping the basement (Fig. 6), and the distribution of rocks susceptible to fracturing (Fig. 9) appear to be responsible for the accumulation of oil on the flanks rather than at the crest of the buried hill.

3. Another assumption is that the most permeable zone in the basement complex lies directly below the unconformity between these rocks and the overlying sediments. The assumption is not generally supported by field evidence as fractures in the upper part of the basement complex above oil accumulations are commonly filled with calcite thus forming a relatively impermeable capping separating basement pools from overlying sediments.

A typical core record across the impermeable zone is that of the Ohio Oil Company's well, Block L2 in the Playa del Rey field (White, 1946, p. 4).

Depth in Feet		
From	*To*	
5,685	5,707	Green-grey hard weathered schist.
5,707	5,770	Blue-grey hard schist.
5,770	5,811	Same with streaks of soft talcy schists.

Depth in Feet		
From	To	
5,811	5,847	Grey and green-grey soft, very talcy schist with a few oil stained streaks.
5,847	5,929	Blue-grey hard schist with talcy streaks.
5,947	6,001	Grey and brownish-red mottled, hard granitoid rock. Red material is possibly hematite. Entire core is slickensided with talcy material in fault planes.
6,016	6,053	Green-grey, oil stained, hard schist.
6,053	6,071	Green and olive brown soft weathered talcy schist.
6,071	6,074	Dark grey and brown oil-stained schist.
6,142	6,219	Green and grey schist fragments.

The original depth of the well was 5,719 feet and both the schist and overlying schist conglomerate were open for production. The well produced at a rate of 125 barrels per day. Subsequent deepening of the well into the softer oil-stained facies of the schist increased its production to almost 2,000 barrels per day.

Other field data indicating an impermeable capping between oil accumulations in the basement and overlying sediments are: (a) depleted oil sands in sediments directly above sizable basement pools (e.g.: Santa Maria Valley field); (b) differences in bottom-hole pressures above and below the barrier (e.g.: Wilmington); and (c) water-saturated sand overlying the basement oil zone (e.g.: Western Gulf Oil Company's Di Giorgio 3, Mountain View field).

4. A capping of impermeable sediments form an integral part of the basement trap depicted in the hypothetical structure in Figure 2. This assumption can not be reconciled with the field evidence where productive sands rest on the basement directly above basement pools (e.g.: Santa Maria Valley, Playa del Rey, and Wilmington fields).

An expectable consequence of the concept illustrated in Figure 2 is that the oil-water interface in the schist reservoir will be found relatively uniform in elevation (Cabeen and Sullwold, 1946, p. 19), with the fractured schist being wet beneath the oil accumulation. This does not occur in fields where sufficient data are available to study oil-water relationships.

The distribution of wet wells in the Edison field is shown in Figure 9. Wells that were wet on testing are on the periphery of the productive area at various depths ranging from 3,200 feet up to 500 feet in sub-sea elevations.

The same "anomalous" relationships between oil and water exist also in the Wilmington field where some basement wells on the crest of the structure were wet on testing in contrast to some structurally lower wells which flowed clean oil on their original completions.

ROLE OF DILATANCY IN MIGRATIONS OF OIL INTO BASEMENT RESERVOIRS

An alternative proposition to the "up-slope" theory of migration was developed as a result of difficulty in reconciling the foregoing assumptions with field observations. According to the alternative hypothesis, migration of oil is due to fracturing of competent basement rocks. Fracturing involves dilatancy which in turn reduces hydrostatic pressures in focal areas of deformation. Pressure gradients are thereby established between the potential reservoir rocks and the over-

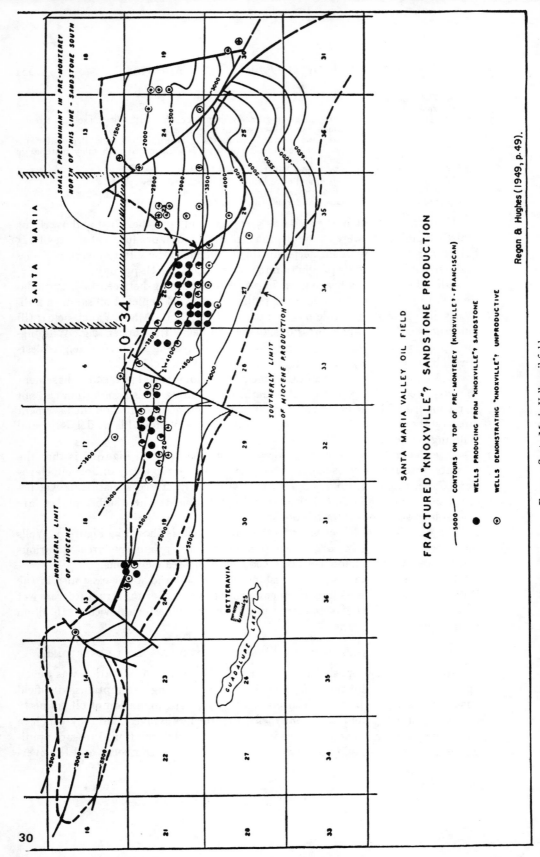

SANTA MARIA VALLEY OIL FIELD

FRACTURED "KNOXVILLE"? SANDSTONE PRODUCTION

—5000— CONTOURS ON TOP OF PRE-MONTEREY (KNOXVILLE? - FRANCISCAN)

● WELLS PRODUCING FROM "KNOXVILLE"? SANDSTONE

◉ WELLS DEMONSTRATING "KNOXVILLE"? UNPRODUCTIVE

Regan & Hughes (1949, p. 49).

FIG. 3.—Santa Maria Valley oil field.

lying source and carrier beds containing oil, gas, and water. Thus, a tendency to "suck in" fluids into the basement is created. Cementation of the upper part of the basement complex is attributed to deposition of calcite from waters associated with oil and gas in the overlying sediments. This is accomplished by release of pressure in the dilated zone thus allowing carbon dioxide to escape from solution

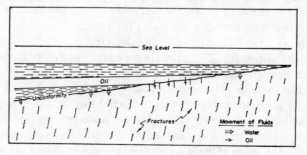

Stage A. Fracturing of metamorphic rocks establishes pressure gradients across the unconformity. Oil, gas, and water move into the dilated basement complex.

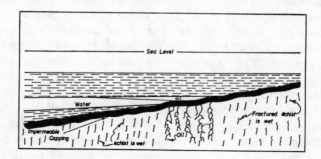

Stage B. Continued fracturing, followed by influent seepage and cementation, has established discrete, cell-like reservoirs below impermeable capping. Fractured schists bordering the reservoirs are wet.

FIG. 4.—Schematic sketches showing evolution of basement oil pool according to dilatancy hypothesis.

with the consequent break-down of the soluble bicarbonate and formation of insoluble calcium carbonate.

The dilatancy concept is not new but its application to this particular problem has not been discussed previously as far as can be determined from an examination of the literature. The most complete statement on the geological role of dilatancy is in Professor Mead's (1925) classical paper in which he uses the term dilatancy for all increases in rock volume due to deformation.

The evolution of a basement oil pool according to the dilatancy hypothesis is depicted by two schematic sketches in Figure 4. The probable sequence of events has been simplified in order to develop the salient features inherent to the hypothesis.

An assemblage of sediments whose basal members buttress against the basement is shown in Stage A (Fig. 4). One member of the assemblage is assumed to be an oil-saturated sand for simplicity in illustration. Source beds of oil undergoing compaction could be substituted for the sand without modifying the concept in any essential respect. The underlying basement complex constitutes part of the trap due either to the absence of significant porosity or to connate water filling existing fractures in the basement complex.

Fracturing of the metamorphic rocks initiated during Stage A (Fig. 4) causes dilatancy of the basement thus establishing pressure gradients across the unconformity. Oil and water move across the unconformity to fill interstices in the fractured rocks—the type of fluid being determined by the fluid content of the sediments in contact with various sectors of the dilated zone. Thus, oil with minor amounts of connate water will enter the basement beneath the oil-saturated sand and water will enter the basement where it is capped by sediments containing only water.

Continued pulsatory tectonic movements finally establish the sizable oil reservoir illustrated in Stage B (Fig. 4). The fractured schists below the oil-bearing sand have become saturated with oil whereas the fractured schists beneath the sediments containing water are wet. Intercommunication of openings is restricted laterally with the result that "anomalous" oil-water relationships are maintained. Consequently, wells on the periphery of the basement oil accumulation will be wet regardless of their elevation on the flanks of the buried hill.

The impermeable capping separating the oil accumulation in the basement complex from the overlying oil sand in Figure 4 was formed by deposition of calcite in fractures in the metamorphic rocks. Deposition of calcite is due indirectly to dilatancy as was noted in the introductory statement of the dilatancy hypothesis.

Thus, wells drilled for the purpose of draining oil in the sand shown in Figure 4 will not drain the underlying reservoir unless penetration of the impermeable capping is effected by accident or design.

TESTING DILATANCY HYPOTHESIS

The dilatancy hypothesis may be tested by deducing geological consequences of the hypothesis and comparing these with observed field relationships. The following geologic relationships are significant insofar as testing is concerned.

1. Oil accumulations in basement rocks should be subjacent to petroliferous sediments, on the assumption that migration of oil in the basement complex is restricted by the lack of free intercommunication between fracture systems. The latter assumption is supported by the "anomalous" oil-water relationships described in the foregoing section.

The field evidence relating to the influence of capping on basement accumulation is not definitive except in the Santa Maria Valley field where Regan and Hughes (1949, p. 49) have shown that accumulation of oil in the basement is subjacent to the basal Miocene producing zone and that it is not present at higher structural positions (Fig. 3) where the non-productive Sisquoc formation of Pliocene age rests on the basement.

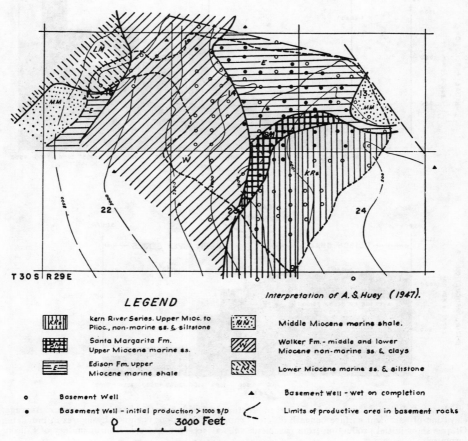

Interpretation of A. S. Huey (1947).

LEGEND

kern River Series. Upper Mioc. to Plioc., non-marine ss. & siltstone	Middle Miocene marine shale.
Santa Margarita Fm. Upper Miocene marine ss.	Walker Fm.- middle and lower Miocene non-marine ss & clays
Edison Fm. upper Miocene marine shale	Lower Miocene marine ss. & siltstone

o Basement Well   ▲ Basement Well - wet on completion

• Basement Well - initial production > 1000 B/D   Limits of productive area in basement rocks

0 ————— 3000 Feet

FIG. 5.—Rock cover on basement complex.

The influence of capping on accumulation in the basement rocks underlying Edison field is susceptible to several interpretations. Figure 5 (Huey, 1947) shows the spatial distribution of the various sedimentary units on the basement surface. Figure 6 shows the relationship of initial production of oil to cap rock. Initial production is plotted on the ordinate, and distance from the Edison shale contact with other sedimentary units is plotted on the abscissa. Basic data for this figure were obtained from Beach and Campbell (1946) and Huey (1947).

The relationships depicted on the graphs may be summarized by stating that most wells completed in the basement beneath the Edison shale had initial productions of more than 1,000 barrels per day, whereas most wells completed in the basement below the Kern series or the Walker formation had much lower initial

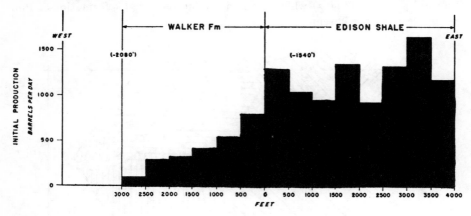

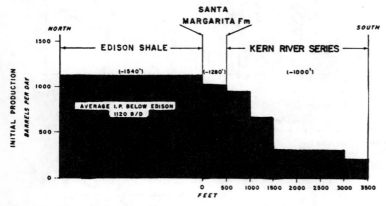

FIG. 6.—Relationship between initial production and rock cover on basement complex. Average elevation of basement surface beneath different capping formations is shown by figures in brackets. Graphs show initial production from basement wells at 500-foot intervals from boundary of Edison formation.

productions. Furthermore, the initial production from wells decreases as distance from the Edison shale member increases.

At least two contrasting interpretations of these relationships are possible. The oil in the basement complex may have originated in the Edison shale and subsequently it was "sucked" into the basement as a result of dilatancy accompanying fracturing. Alternatively, the Edison shale may have been an impermeable barrier to the upward migration of oil from the basement into overlying sediments and thus have constituted a part of the trap as suggested in Figure 2 (Cabeen and Sullwold, 1946, p. 25).

Acceptance of the latter interpretation is difficult to reconcile with the oil accumulations beneath sands comprising the basal members of other sedimentary units capping the basement. Thus, the possibility that oil in the basement complex originated in the Edison shale is favored by the writer.

2. Oil-bearing fractures in the basement complex should be tectonic in origin.

The field evidence concerning origin of the fracture systems is definitive. They extend to depths well beyond the expectable range of openings due to superficial processes associated with erosion and weathering during the geologic epochs when the erosion surface at the top of the basement was being sculptured.

Thus, as would be expected, the initial rate of production of oil is related to the depth of penetration—the greater the penetration, the higher the initial production from basement wells in all fields where sufficient data are available to evaluate this relationship. This is shown graphically in Figure 7.

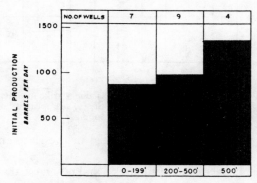

FIG. 7.—Relationship between initial production and depth of penetration into basement in that part of field where rocks of "transition zone" are overlain by Edison shale.

The influence of factors other than the amount of penetration was minimized in the compilation of production data by selecting twenty wells in the Edison field that were completed beneath the Edison shale capping (Fig. 5) in basement rocks having broadly similar lithologic characteristics (transition zone lithologic assemblage, Figs. 8 and 9).

3. Oil-bearing fractures should have been formed after the deposition of superjacent sediments.

Fracture systems of widely different ages are in the basement complex. Consequently, the determination of the age of oil-bearing fractures is difficult. Old fracture systems that were healed by deposition of minerals from circulating ground waters have been reopened by later movements and oil has entered the fractures. This sequence of events is apparent in some cores from the Edison field where impregnations of oil are in the medial sections of seams of calcite.

Despite the obvious difficulties in establishing the age of fractures, there is indirect field evidence indicating that ground preparation for the later accumulation of oil in the Edison field is related to a structure affecting both the basement complex and overlying sediments. Figure 8 shows average initial productions from wells completed in three different lithologic units (Fig. 9) at distances up to one mile from the Edison fault system along the northern margin of the field.

Average initial productions from wells completed in the different lithologic units were determined for each interval having a width of 1,000 feet and then average initial productions were plotted on the graph at 500 feet, 1,500 feet, *et cetera.*

Figure 8 shows that wells within a distance of 1,000 feet of the Edison fault had the highest initial productions and that initial productions decrease in a

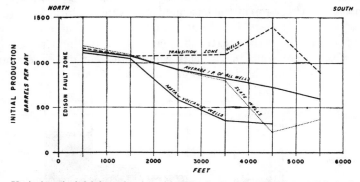

Fig. 8.—Variations in initial productions of wells along strike of different lithologic units. Controlling factors: (1) fracturing adjacent to Edison fault; (2) susceptibility of different lithologic units to fracturing; and (3) distribution of Edison shale capping.

southwestward direction in wells completed in the slate and the meta-volcanic lithologic units.

The relationships depicted in Figures 8 and 9 suggest that brecciation of basement rocks adjacent to the Edison fault system was an important factor in ground preparation for the oil reservoir in Edison field. Movement along this fault system was periodic during the deposition of sediments of Miocene and lower Pliocene age as is indicated by stratigraphic displacements increasing with the age of the sediments offset by faulting.

Little information is available on the age of oil-bearing fractures in the basement complex underlying the Wilmington, El Segundo, Playa del Rey, and Santa Maria Valley fields. Winterburn (1943, p. 304) states that the principal faults in the Wilmington field were active during deposition of the Repetto formation of the lower Pliocene. An investigation of the relationship between productivity of basement wells and faulting was not feasible in the Wilmington field because of completion practices—multiple completions involving both basement rocks and the overlying sands were common in the early stages of development of the field.

4. Some lithologic control of fracturing should be evident if secondary openings in the reservoir rocks are tectonic in origin.

This occurs in fields where the distribution of different lithologic units comprising the basement complex is known.

Figures 8 and 10 show the relationship between lithologic character and initial production in Edison field. Differences in initial production from the various lithologic units are probably related to contrasting physical characteristics of rocks comprising the units.

For example, the most favorable lithologic unit insofar as oil accumulation is concerned is the transition zone between meta-volcanics and meta-sediments. It

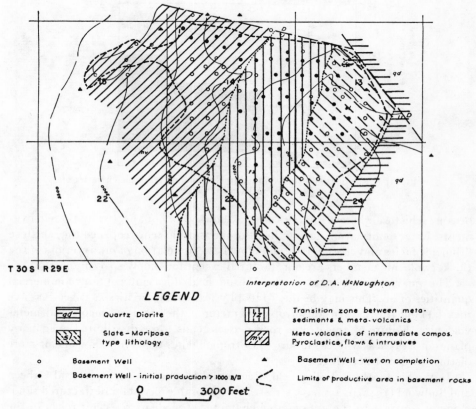

Interpretation of D. A. McNaughton

LEGEND

⊟qd⊟	Quartz Diorite	⊞	Transition zone between meta-sediments & meta-volcanics
▨s▨	Slate – Mariposa type lithology	▨mv▨	Meta-volcanics of intermediate compos. Pyroclastics, flows & intrusives
○	Basement Well	▲	Basement Well - wet on completion
●	Basement Well - initial production > 1000 B/D	⌇	Limits of productive area in basement rocks

0   3000 Feet

Fig. 9.—Basement rocks at Edison field.

includes thin relatively brittle volcanic flows intercalated with tuffaceous sediments and argillite. These rocks were closely folded in a deep-seated environment during the orogeny preceding intrusion of the Sierra Nevadan batholithic rocks (quartz-diorite pluton in Edison field) in the late Jurassic. Slippage along bedding planes and contacts during folding caused fracturing of thin brittle members, and flowage and partial recrystallization of incompetent members. Secondary openings in the rocks resulting from these movements were partially if not completely healed by deposition of secondary minerals deposited from hydrothermal solutions emanating from the quartz-diorite pluton. Nevertheless they constituted planes of weakness in the rocks which were susceptible to reopening in near-surface environments during Miocene and Pliocene crustal movements.

Similarly, folding of the meta-volcanic lithologic unit was accompanied by slippage between massive volcanic members. Contacts between these members

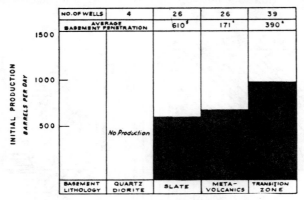

FIG. 10.—Relationship between initial production and lithologic character of basement rocks in Edison field.

became schistose whereas interiors of thick flows were not affected by these movements. Later deformation during the Miocene and Pliocene reopened the schistose lithologic facies and created reservoirs for the accumulation of oil. Interiors of the thicker volcanic members are not fractured except for widely spaced joints.

The quartz-diorite mass in the Edison field does not contain commercial quantities of oil. This may be due to its physical characteristics as it is a massive rock lacking the marked anisotropic character of the meta-sediments and meta-volcanics. Failure of anisotropic rock masses tends to be distributed over many planes of weakness whereas failure of isotropic bodies tends to follow one or more distinct lines of fracturing.

In the Santa Maria Valley field, Regan and Hughes (1949, p. 51) and Cabeen and Sullwold (1946, p. 25) have stressed the favorable character of fractured sandstone in contrast to the unfavorable character of shale as reservoir rocks in the basement.

The relationships between basement lithologic character and fracturing in the Wilmington, El Segundo, and Playa del Rey fields are not known as insufficient cores are available for mapping purposes.

APPLICATIONS OF DILATANCY HYPOTHESIS

Exploration for basement oil accumulations guided by the dilatancy hypothesis will be focussed on areas where (a) petroliferous sediments (carrier and source beds) buttress against basement rocks, (b) competent members are in the metamorphic assemblage comprising the basement complex, and (c) both basement complex and overlying sediments are affected by a fault or a fold whose position in the basement complex can be predicted within at least 1,000 feet. Thus, targets for exploration may be established if the hypothesis is correct.

CONCLUSIONS

The primary purpose of this paper is to stimulate geological discussion of basement oil accumulations. The dilatancy hypothesis has not been advanced as a panacea for all of the geological woes associated with exploration for oil accumulations in basement rocks. They will always be difficult targets to predict regardless of whether they occur in secondary openings of tectonic origin or in regolithic débris capping the basement. Thus, they pose a challenge to geologists engaged in exploratory work.

REFERENCES

BARTON, C. L. (1931), "A Report of Playa del Rey Oil Field, Los Angeles County, Calif.," *California Oil Fields*, Vol. 17, No. 2 (October, November, December), pp. 5–15; 5 pls., incl. map and sections.

BEACH, J. H., AND CAMPBELL, H. (1946), "Geology of Basement Complex, Edison Oil Field, Kern County, California" (abstract), *Bull. Amer. Assoc. Petrol. Geol.*, Vol. 30, No. 12, p. 2090.

BROWN, A. B., AND KEW, W. S. W. (1932), "Occurrence of Oil in Metamorphic Rocks of San Gabriel Mountains, Los Angeles County, California," *ibid.*, Vol. 16, No. 8, pp. 777–85.

CABEEN, R. C., AND SULLWOLD, H. H. (1946), "California Basement Production Possibilities," *Oil Weekly* (June 22), pp. 17–25.

HUEY, A. S. (1947), "Geology of Basement Complex, Edison Field, California," paper read at Amer. Assoc. of Petrol. Geol. 32d Annual Meeting, Los Angeles, California.

MAY, J. C., AND HEWITT, R. L. (1948), "The Basement Complex in Well Samples from the Sacramento and San Joaquin Valleys, California," *California Jour. Mines and Geol.*, Vol. 44, No. 2, pp. 129–58.

MEAD, W. J. (1925), "The Geologic Rôle of Dilatancy," *Jour. Geol.*, Vol. 33, No. 7, pp. 685–98.

REGAN, LOUIS J., AND HUGHES, ADEN W. (1949), "Fractured Reservoirs of Santa Maria District, California," *Bull. Amer. Assoc. Petrol. Geol.*, Vol. 33, No. 1 (January), pp. 32–51.

WHITE, J. L. (1946), "The Schist Surface of the Western Los Angeles Basin," *California Dept. Nat. Resources, Div. Oil and Gas, 32d Ann. Rep.*, pp. 3–13.

WINTERBURN, R. (1943), "Wilmington Oil Field," *Div. Mines Bull. 118*, pp. 301–5.

Copyright 1953 by The American Association
of Petroleum Geologists
BULLETIN OF THE AMERICAN ASSOCIATION OF PETROLEUM GEOLOGISTS
VOL. 37, NO. 2 (FEBRUARY, 1953), PP. 250-265, 10 FIGS.

FRACTURING IN SPRABERRY RESERVOIR, WEST TEXAS[1]

WALTER M. WILKINSON[2]
Midland, Texas

ABSTRACT

The Spraberry formation of West Texas is developed in the lower Leonard of middle Permian, restricted in most part to the Midland basin. The main producing structure is a fractured permeability trap on a homoclinal fold. This homogeneous mass is undifferentiated except as to alternate layers of sands, siltstones, shales, and limestones, deposited in a deep basin under stagnant conditions with hydrocarbons formed throughout the 1,000 feet of sedimentary rocks.

Fractures were created by tensional forces after induration, probably during post-Leonard time. With storage of the oil reservoir in the sandstone matrix, the fractures serve as "feeder lines" to conduct the oil to the bore hole. Without these fractures commercial production would be from a seemingly "too-tight" reservoir rock.

The producing area of the Spraberry formation is a "fairway" 150 miles long and 50 miles wide at an average depth of 6,800 feet. The main area, however, is 50 miles long, with width ranging from a few miles to 48 miles, thus creating a triangle of 488,000 proved and semi-proved productive acres.

INTRODUCTION

The Spraberry trend (Fig. 1) is distributed throughout the main part of the Midland basin, a geological province of the Permian basin. The major part of the trend extends north and south 150 miles and attains maximum width of nearly 75 miles.

Physiographically, the trend area is in the region between the south end of the Llano Estacado and the north part of the Edwards Plateau. The northern area is mostly sand and sandy soils, while the southern area contains tighter soils which are predominantly clay loams. The topography is marked by low hills, draws, and dry lakes without drainage. Recent sands and gravels, Tertiary gravels, Cretaceous limestones and sandstones, and Triassic redbeds are exposed with no recognized, but not necessarily unrecognizable, surface expression of the subsurface structures. The climate is semi-arid with a mean annual rainfall of 18–20 inches. The surface cover is mainly prairie grasses and mesquite.

HISTORY OF DEVELOPMENT

In 1944 a dry pre-San Andres test was drilled by the Seaboard Oil Company on the Abner J. Spraberry farm of east-central Dawson County. During the drilling of this well, a sandstone was noticed to have a slight showing of oil or gas and was locally called the "Spraberry sandstone." No completion was attempted in this sandstone, and no significance was attached to the possible reservoir rock until the Seaboard Oil Company's Lee well No. 2-D was drilled in late 1948 to the Ellenburger formation. During the drilling of this well, a showing of oil was noted at approximately 7,000 feet and a decision was made to test this zone. The Lee No. 2 was subsequently completed as a producer with flowing potential of 319

---

[1] Read before the Association at Los Angeles, March 28, 1952. Manuscript received, July 1, 1952.

[2] District geologist, Sohio Petroleum Company.

250

40

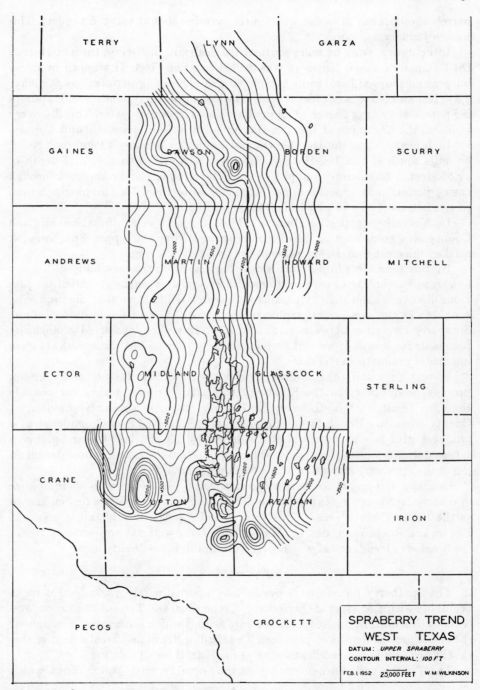

FIG. I.—Spraberry trend, West Texas. Structure map contoured on top of Spraberry formation.

barrels of oil per day after 640-quart nitroglycerine shot at 6,455–6,535 feet. This was on January 22, 1949.

In February, 1949, 65 miles south of the Seaboard discovery, the Tex-Harvey Oil Company's Floyd No. 1–16 was drilled to 12,063 feet. It stopped in Ellenburger and plugged back to be completed through casing perforations at 7,865–7,875 feet and 8,045–8,055 feet opposite the lower Spraberry sandstone. Its pumping potential was 135 barrels of oil per day, plus 13 per cent water. This discovery well was the beginning of the Tex-Harvey oil field in eastern Midland County.

In January, 1950, the Humble Oil and Refining Company's Pembrook No. 1, 30 miles south of the Tex-Harvey discovery, was drilled to the total depth of 12,660 feet in Ellenburger and plugged back to 7,169 feet and completed through casing perforations opposite upper Spraberry sandstone. Its pumping potential was 34 barrels of oil and 4 barrels of water per day.

In November, 1950, the Humble's Midkiff No. 1 in southeastern Midland County was completed as small pumping well from the upper Spraberry; it marked the most western extension of the trend at that time.

By this time some importance was being placed on the possibility of "shoreline trend" according to conversations with several area geologists. All of the previous discoveries and their extensions were correlated to show that the producing zone was in the same stratigraphic reservoir rock. Subsea limits on top of the Spraberry formation between 4,200 and 4,500 feet were considered geologically justifiable for a rush of wildcat activity in early 1951. All of these wildcats were important productive extensions.

During 1951 and 1952 additional wildcats were drilled structurally updip, proving productivity in the Spraberry formation without regard for possible shoreline trend, with a discovery in January, 1952, in southwestern Sterling County when the Honolulu Oil Corporation's Sugg No. 1 was completed as a producer with flowing potential of 349 barrels of oil per day. Upper Spraberry subsea datum was 2,450 feet, or 2,100 feet updip from the lowest subsea datum on top of the Spraberry formation which was then productive.

On May 1, 1952, there were 1,630 completed producing wells in the entire Spraberry trend, and 1,558 producing wells in the main productive region, known as the "Four-County area." This area includes parts of Midland, Glasscock, Upton, and Reagan counties. On May 1, 1952, there were 243 active rotary rigs in the Spraberry trend, 208 of which were busy in the Four-County area.

STRATIGRAPHY

The Spraberry formation is overlain by approximately 7,000 feet of rocks beginning with 1,600 feet of Quaternary, Cretaceous, and Triassic sandstones and redbeds. Included in this sequence is a caliche unit which is prevalent throughout the entire area at depths of 5–10 feet. The shallow depth and areal extent of this caliche bed provide an excellent source of road metal for oil-field use.

The Triassic rests unconformably on the upper Permian Ochoa series, which consists of 1,000 feet, more or less, of redbeds, halite, polyhalite, and anhydrite.

Between the Ochoa series and the top of the Spraberry are approximately 4,700 feet of rocks belonging in the Guadalupe series and the upper part of the Leonard series. The upper 1,700 feet are interbedded dolomite and clastic beds and the lower 3,000 feet are interbedded dolomite and black shale with several sandstone beds prominent locally. It should be mentioned here that the black fissile shale and the thin dolomite beds of the Clear Fork group which directly overlies the Spraberry are fractured similarly to the Spraberry rocks and with the same lithologic appearance.

The Spraberry formation (Fig. 2) is approximately 1,000 feet thick and is generally composed of 852 feet of black shales and silty shales, 131 feet of silt-stones, and 5 feet of thin-bedded limestones, or dolomites. This formation belongs to the lower Leonard[3] and rests conformably on the Wolfcamp. The mass of rocks can be separated into three distinct and correlative units, which are classified as upper Spraberry, middle Spraberry, and lower Spraberry, with approximately 330 feet of rocks assigned to each unit.

Some of the characteristics of the common rock types found in the Spraberry formation are listed and have been determined by petrographic descriptions from the Sohio Petroleum Company's Mary V. Bryans No. 1, Sec. 12, Block 37, T. 5 S., Glasscock County. The texture and mineralogical character of the major constituents have been considered more important than the identification of the heavy minerals and organic remains.

Siltstones.—The grain-size description indicates that the major percentage of grains falls in the silt-size range with approximately 60 per cent between the grade limits of 0.03–0.06 millimeter. The grains range from angular to very angular, and the sorting from fair to poor. Primarily, the cementing agent is dolomite with some silica. All gradations are present from a nearly pure siltstone to those containing nearly 50 per cent dolomite. In association with the siltstones are some minor constituents such as pyrite, mica, and plagioclase feldspars.

Dolomites.—The dolomites vary in texture from fine to crystalline, the fine crystals being primary. It was observed that the dolomites generally corrode quartz grains.

Shales.—Several types of shales have been encountered, such as massive, blocky type as well as the commonly found fissile, brittle type. Most of the shales have been classified as carbonaceous, but in all petrographic slides examined, the shales were found to be silty and very ferruginous.

In all of the wells examined it was observed that the siltstones, dolomites, and shales were very similar areally and vertically in texture and mineral character. Certain distinctions can be made by electrical-log study (Fig. 2) to separate the upper Spraberry from the middle Spraberry, as well as to separate the middle Spraberry from the lower Spraberry. However, there does not appear to be any

---

[3] Lamar McLennan, Jr., and H. Waring Bradley, "Spraberry and Dean Sandstones of West Texas," *Bull. Amer. Assoc. Petrol. Geol.*, Vol. 35, No. 4 (April, 1951), p. 899.

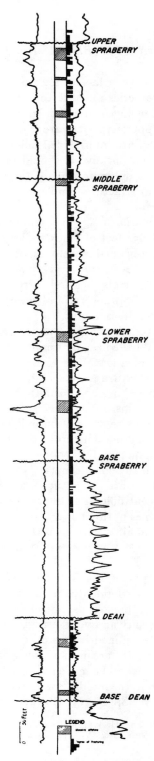

FIG. 2.—Generalized columnar section of Spraberry formation. Diagonal shading, massive siltstone. Solid black, degree of fracturing.

easily identifiable characteristic which differentiates the clastic content in the three units, one from the other, if determination is made by binocular or petrographic microscope.

A generalized lithological description of each of the three units follows.

Upper Spraberry.—323 feet of sedimentary rocks composed of approximately 45 feet of siltstone, 3 feet of limestone, and 275 feet of black shale or gradations thereof, which include an increase in silt content.

Middle Spraberry.—361 feet of sedimentary rocks composed of approximately 26 feet of siltstone found in very thin beds and not of the more massive type similar to upper Spraberry, one foot of limestone, and 334 feet of black shale and silty shale.

Lower Spraberry.—304 feet of sedimentary rocks composed of approximately 60 feet of siltstone with the more massive type found near the base of the section, 1 foot of limestone, and 243 feet of black shale and silty shale.

Specific attention has been paid during the preparation of this paper to the type of deposits and original environment in the Four-County area. The remaining part of the Spraberry trend has a general similarity, but, with the exception of Spraberry Deep pool, commercial production of major significance has not been developed. The writer believes that the Midland basin as a whole had partly restricted water circulation during Spraberry time. The Four-County area was notably different from other parts of the basin, however, in that the circulation was so restricted as to cause toxicity in the bottom waters, with resulting sediments of euxinic facies. This region appears to have been a partly isolated part of the broad shallow basin with confinement of organisms to the top waters and with the deeper bottom waters oxygen-deficient and lacking in benthonic life, thus indicating true euxinic conditions.

The following is a general long-range correlation of texture. A well was examined in northwestern Lynn County (Anderson-Prichard's White No. 1, Sec. 154, HS&WT) in the northern end of the Spraberry trend. The upper Spraberry exhibits gross mineral character and texture nearly identical with rocks examined farther south in the Four-County area. The samples examined do not contain any appreciable clastics in the sand-size range. East of the Four-County area, there is an increase in the median diameter of grain size. In the Honolulu Oil Corporation's Sugg No. 1 of southwestern Sterling County, the texture is considerably coarser than other sections examined farther west. The sample of upper Spraberry in this well is very fine-grained sandstone with the largest percentage of grains in the upper limit of a very fine sand classification.

The feldspar materials examined contain no enlarged feldspar grains, but consist of a detrital core with a secondary rim. All shale samples examined are silty, even the black, carbonaceous types; nearly all specimens contain at least 10 per cent silt-sized material. Intermixed with the shale, here and there, is a bit of argillaceous material, which consists of predominantly fine, micaceous shreds (sericite) and kaolin.

ACCUMULATION OF OIL

As previously mentioned, a representation of "shoreline trend" was developed during the early part of Spraberry exploration. Long north and south extensions restricted production to subsea limits on top of Spraberry formation between 4,200 feet and 4,500 feet. At this time, however, oil is produced between subsea limits

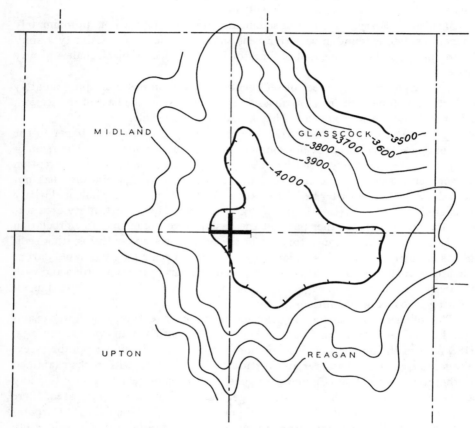

FIG. 3.—Structure map of Four-County area contoured on top of Dean sand, with regional dip eliminated.

of the Spraberry formation of 2,450–4,550 feet (Fig. 1). A sufficient number of dry holes were drilled in 1951 to partially delineate the producing area as approximately triangular. Only the western side of the triangle follows datum and roughly established water table, whereas the other two sides of the triangle follow no established geological pattern. It is evident, therefore, that some controlling factor is present for the accumulation of Spraberry oil in its present position other than a "shoreline trend," or facies change. With this idea in mind, a map (Fig. 3) of the Four-County area was prepared to eliminate regional dip and approximate the probable topography of the sea floor during early Spraberry deposition.

During the preparation of paleogeographic maps, several mapping techniques are used, Figure 3 illustrating only one. Such a map merely aids in the technique of visually restoring rocks to their relative original position of deposition. To show the topographic features at the end of Wolfcamp time, the top of Dean sand was selected to represent the near close of Wolfcamp time. The stratigraphic position of the Dean sandstone relative to the Spraberry is shown in Figure 2. The Dean sand has been identified as Wolfcamp in age, but used as datum point since base of Spraberry—top Dean interval is consistent in thickness. With the Dean data, the present subsurface was rotated vertically from an arbitrary hinge line that approximated the western edge of the Eastern platform. The map is, of necessity, a rough semblance of the ideal, but it does show a gentle, closed low area that closely follows the areal distribution of the major productive region.

During Permian time the Midland basin was a mildly negative area receiving sediments from far-removed emergent lands, and these sediments were deposited in deep, quiet waters. The basin was not necessarily restricted in the environmental sense. Circulation was open at the north and south; however, isolated low areas on the main basin floor were locally restricted in circulation. In such closed low areas the sea water could maintain its volume but would stagnate. Surface waters were favorable for supporting many planktonic forms in its pelagic realm. Contribution of organic material from these surface forms, as well as floating débris, could be preserved on basin floors where oxygen was deficient and hydrogen sulphide content was high. Rich organic muds were thus formed, as evidenced by much bituminous matter and in this type of environment, anaerobic bacteria would be encouraged to break down the organic material into simpler and more basic hydrocarbons. The writer believes that such conditions did exist in the gentle, closed low area shown in Figure 3 at the beginning of Spraberry deposition.

This writer believes that the Midland basin was relatively non-toxic in most part, and yet it was probably pitted with several low restricted spots of varying depths. In these low areas would be the richer accumulation of organic muds. Present distribution of oil or gas points to these areas individually, and tends to show that the generation of oil was localized to a great degree.

A lithofacies map (Fig. 4) of the Four-County area further substantiates the reconstruction of the geological features at the close of Wolfcamp time. Northeast of the Spraberry trend non-clastics are dominant with a gradual facies change toward the productive area, where approximately 87 per cent of the Spraberry is shale. West of the producing area a rather abrupt shale to limestone facies change occurs, with several wells on the west side of the present structural highs containing 85 per cent non-clastics. This probably means that during the time of Spraberry deposition, the west side of the Midland basin was elevated sufficiently to provide an environment of warm, shallow waters, in which carbonates were precipitated and deposited. Superimposed on the lithofacies map are isopach contours representing the thickness of Spraberry-type rocks after total deposition

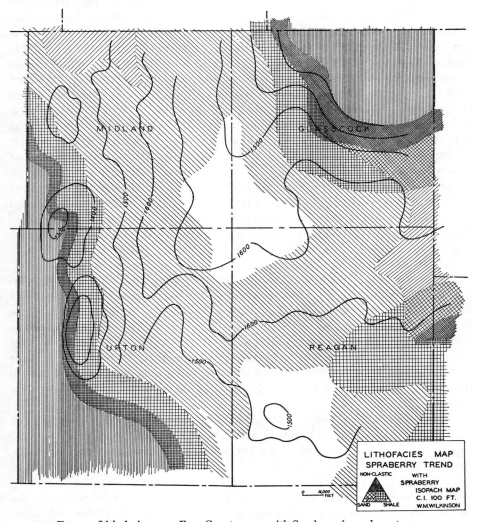

FIG. 4.—Lithofacies map, Four-County area, with Spraberry isopach contours.

and after accounting for all structural movements. Included in the interval thickness is the 1,000 feet of Spraberry section plus approximately 500 feet of upper Wolfcamp rocks to the base of the Dean sand. The area of thickest sedimentation is found in the south-central part of Glasscock County and the north-central part of Reagan County, and appears to be associated with the original low area, interrupted only by slight structural growth around the common corners of the Four-County area.

The accumulation of Spraberry oil was probably directly associated with a relatively deep, closed basin since present structural position is no determining factor for the presence or absence of a productive reservoir, nor is a lithologic change a determining factor for the presence or absence of oil. Specific reference is made to an arbitrary line representing the south limb of the triangular area trending eastward through central Reagan County. The rocks north of this arbitrary line are generally the same as the rocks south of this line; however, productive reservoirs are found at the north and increase in productivity toward the center of the triangular area.

FRACTURES

Even though many diagnostic characteristics of jointing are present, there is no evidence of vertical displacement, but there is a fairly well defined pattern of

FIG. 5.—Core showing incipient or latent vertical fractures. Diameter of core, 3½ inches.

fracture trend, both major and minor, which may establish a fracture, or joint system.

Fractures have been assigned a set of arbitrary indices by the writer: (1) latent fractures (Fig. 5); (2) single vertical or oblique fractures, discontinuous for a relatively short distance; (3) single vertical fracture, extending for entire length of lithologic unit (Fig. 6); (4) single vertical fractures parallel with each other; and (5) vertical fractures parallel, intersected by oblique or vertical cross fractures (Fig. 7).

The most common type of fracture with the greatest continuous vertical extent is found in the black, brittle shales and in the varved, sandy shales. Oblique fracturing occurs by far the most commonly in the silts but, where present in the shales, occurs alone and is discontinuous up to 18 inches in length. The oblique fracturing has a tendency to assume a position of shattering with the end

Fig. 6.—Core of shale and siltstone showing single vertical fracture extending across lithologic units. Diameter of core, 3½ inches.

Fig. 7.—Core showing combination of fracture types. Diameter of core, 3½ inches.

result representing a pattern of shingles. This type of shattering has been classified as imbricate. It is interesting to note that the oblique fracturing on one side of the core generally has a compensating latent fracture on the opposite side of the core, but at an opposing angle. The writer suggests that the shattering effect produced in the sands and silts could be a result of compositional difference, as compared with shales, with the sands and silts having reached an elastic limit, and then shattering to relieve the tensional pressure when exposed by the bore hole.

All fractures tend to be centered in the cores, particularly where only single vertical fractures are found. It would be an unlikely geological phenomenon that fractures would fall naturally in a vertical plane within bore-hole deviation tolerances. Several hypotheses can be assumed for this phenomenon. It may be that fracturing is artificial to some degree and is created by stress relief along prede-termined lines of weakness, or planes of microscopic fracturing. It may be that the lines of fracturing are so closely spaced that the wandering action of the drill bit or core head would find the line of weakness and follow it, as in steeply dipping beds in areas of thrust faulting. However, mineralization of fracture planes (Fig. 8), presence of lost circulation materials, and cement within fractures after coring certainly prove that fissures are open, even though the presence of cement is highly overrated. Abnormal injection pressures would aid in forcing foreign ma-terials into the earth opening. Some operators have introduced carnotite into the cement slurry before casing is cemented above the Spraberry formation. The gamma-ray log showed no evidence of cement below casing point.

Pore space in the fractures before coring has been partly determined by the

Fig. 8.—Cores showing intersecting vertical fractures and mineralization. Diameter of core, $2\frac{7}{8}$ inches.

use of micrometer gauge. An over-all average for a single vertical fracture is 0.002 inch. Certainly capillary and sub-capillary openings can not be measured, even though they have been proved to be present. These immeasurable openings play a tremendous role in the movement of fluids. Since shales constitute 87 per cent of the Spraberry rocks and since the presence of sub-capillary openings is es-tablished, the shales are apt to be reservoir rocks in part. Oil has been retorted from some samples to prove the presence of hydrocarbons within the shale matrix.

The matrix siltstone which serves as the main reservoir rock has an average permeability of 0.50 millidarcy and an average porosity of 8 per cent. With this type of permeability in the reservoir rock, it becomes obvious that the fractures serve mainly as "feeder lines" to conduct oil to the well bore. In very few places has commercial production been developed without a rupturing process to create more fracture permeability channels.

The genesis of Spraberry fracturing can be attributed to two forces: (1) non-directional reduction in volume, and (2) regional tensions created by basinward subsidence. These two changes occurred independently of each other, even though the latter was associated with structural growth from deeper tectonic forces.

Reduction in volume by the removal of interstitial waters is inherent in lithification of muds to form shales. Shrinkage is possible by changes in the various clay minerals making up the shales. The process necessary for creating fissility of shales results in microscopic lateral openings. All of these processes resulted in shrinkage cracks, and after sufficient lithification had taken place for the rocks to become brittle, latent fracturing took place as the volume changed, but resulted only in microscopic lines of weakness without definite direction or trend.

Tests have been made in Sohio laboratories of the different rocks of the Spraberry formation and prove the presence of sub-capillary openings. Fracture plane faces were washed with lubricating gasoline for sufficient length of time to permit absorption of the lubricating material into any sub-capillary openings that might

FIG. 9.—Core showing shrinkage cracks in Spraberry shale. Diameter of core, 3½ inches.

be present. Exposure to air and heat provided rapid evaporation from the fracture plane. In a very short time the fracture face was free of gasoline, but a comprehensive system of shrinkage cracks was proved by the remaining lubricating material in the cracks themselves (Fig. 9). Experiments have shown that this system is present only in the shales. There is very minor evidence of latent fracturing in the siltstones. The presence of the shrinkage crack system in the shales further proves the non-directional reduction in volume.

Core orientation tests have been made in several wells in the Four-County area, and a major fracture trend has been indicated to have a general N. 25° E. direction, with a more poorly developed set of cross fractures normal to the main trend. These wells are: (1) Sohio's TXL "A" No. 6, Sec. 35, Block 37, T. 4 S., Glasscock County; (2) Sohio's Davenport "B" No. 1, Sec. 2, Block 37, T. 5 S., Glasscock County; and (3) Sohio's Bernstein No. 1, Sec. 5, Block N, Upton

County. An initial potential map (Fig. 10) prepared of the Tex-Harvey pool has indicated the same general north-northeast direction. On the initial potential map the black coloring represents the wells of highest productivity and further indicates the very close proximity to better developed fracture system.

The writer suggests that possibly the system of open fractures shown by potential and productivity tests and core orientation tests was created by the effects of regional tension as the basin subsided after Leonard time. The gentle subsidence would stretch the pre-existent mass of rocks from the buttressing positive element of the Eastern platform. This stretching would have a tendency to create the greatest amount of rock rupturing in the area closest to the deepest part of down-warping, but still would be confined to those rocks of highest shale content. Tectonic movements near the end of Permian time re-uplifted certain areas such as the Reagan anticline and the Wilshire-Pegasus fold (Fig. 1). The Reagan anticline has a N. 45° W. direction, and the Wilshire-Pegasus fold has a northerly direction. The down-warping of the basin, combined with the structural growth and uplifting of the two anticlinal folds, would have a tendency to create certain torsional effects on the mass of rocks nearest the area of movement. The torsional and tensional effects of these combined movements on the Spraberry rocks would be apt to produce northwest to west nosings, and any such movement, however slight, would tend further to rupture and connect the ancient lines of weakness.

PRODUCTION STATISTICS AND SUMMARY

Since the date of the discovery well in the Spraberry Deep pool, and since the time of completion of the original Spraberry well in the Tex-Harvey field, an ever-increasing tempo of activity has confronted the oil industry in the Spraberry trend. The greatest concentration of drilling activity and completions has been confined, for the most part, to parts of the Four-County area. It is not important to quote individual month or year statistics, but rather, comparison can be made by quoting widely separated statistics.

In October, 1951, within the Four-County area there were 531 completed wells and 241 drilling operations. Outside of this area there were 72 completed wells with 5 drilling operations.

During the month of April, 1952, there were 1,558 producing wells within the Four-County area, an average of 176 completions per month. During this same month 2,744,156 barrels of oil were produced, which represents 77 per cent of the assigned allowable.

On February 1, 1952, there were 766 rotary drilling units in operation in the Permian basin areas of West Texas and southeastern New Mexico. The Four-County area had 315 of those operations, which represents 41 per cent of the actual drilling rigs in the Permian basin.

With an activity so concentrated, there is a constantly changing picture day by day, and any production charts of to-day would be obsolete tomorrow. It is expected that approximately 488,000 acres will be proved productive at the con-

FIG. 10.—Initial potential map of Tex-Harvey field.

clusion of developmental drilling within the Four-County area. Outside areas appear to be marginal at present and yet areas favored by optimum accumulation of oil and associated with greater fracturing would be favored for production unknown at this time.

Truthfully, it may be said that the Spraberry is a unique reservoir. It is fabulous in areal extent, puzzling with its production problems, and baffling with its geological phenomena. Here in an area devoid of typical folded traps, and from a mass of rocks that would be classified as non-commercial under average conditions, flowed the aforementioned 2,744,156 barrels of oil from 1,558 wells during the month of April, primarily the result of fractures.

BULLETIN OF THE AMERICAN ASSOCIATION OF PETROLEUM GEOLOGISTS
VOL. 41, NO. 8 (AUGUST, 1957), PP. 1748-1759, 10 FIGS.

DEVELOPMENT OF FRACTURE ANALYSIS AS EXPLORATION METHOD[1]

P. H. BLANCHET[2]
Calgary, Alberta, Canada

ABSTRACT

Fracture analysis is based on the premise that the earth's crust is systematically fractured and that this fracturing can be recognized and analyzed on aerial photographs. This fracturing is due to external forces which have been active throughout geologic time. Four sets of fractures, arranged in two groups of two, the orientation of which is regionally systematic, have been found in all areas studied. The local deviations in the statistical mean direction of the fracture sets, when measured and analyzed, are found to be related to structure or stratigraphic anomalies.

The crust of the earth is fractured. It is abundantly and systematically fractured. And it is fractured in four principal directions. Although in part this fracturing may be due to internal stresses, it is the external stresses, which have been active throughout geologic time and which continue to act upon the crust today, that are considered to be largely responsible. Several principal groups of such external stresses are recognized. One of the more important is the terrestrial or earth tides. These tides, due to the gravitational effects of the moon and sun, act upon the solid crust itself. A second important group is related to the changes in the radial acceleration of the earth along its radius vector; a third, to the gradual decrease in the earth's rate of rotation, considered to be due to tidal friction. This concept that the earth's crust is abundantly and systematically fractured is the basic premise on which is built the analytical exploration technique known as fracture analysis.

The objective of fracture analysis is to locate shallow to deep-seated structures and/or stratigraphic anomalies. This is accomplished by mathematically analysing the fractures present in an area and discernible on aerial photographs.

Before intelligently discussing fracture analysis itself, it is essential that certain terms be defined and a broader statement of the basic concept be given.

Firstly, by the term *fracture*, is meant the generally abundant, natural lineations discernible on aerial photographs. Fractures express themselves at the surface in many different ways and in a great variety of forms, dependent, to a large extent, on the type of soil or overburden, on the vegetational cover, and on the drainage and ground-water conditions prevailing in the area. Common to all is their marked rectilinear character which makes them apparent in an otherwise curvilinear to amorphous environment—the normal habit of nature. Only a small percentage of the mappable fractures present on an aerial photograph are obvious or even readily apparent. For this reason, it is extremely difficult to find suitable photographs from which to make half-tone plates to satisfactorily illustrate fractures. Figure 1, however, shows some reasonably apparent fractures

[1] Read before the Rocky Mountain Section of the Association at Salt Lake City, February 26, 1957. Manuscript received, May 27, 1957.

[2] Blanchet and Associates, Ltd.

1748

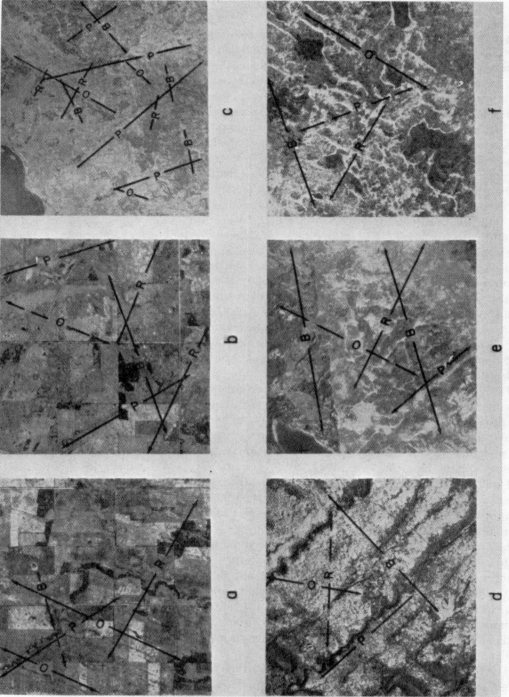

FIG. 1.—Surface expression of fractures in Plains region, in various parts of Alberta. Black arrows are not annotations for any particular fracture. They draw attention to the four distinctive directions of fractures apparent on each photograph.

in six areas in various parts of Alberta. These are sedimentary areas, in which the unconsolidated overburden ranges from a few feet to several hundred feet in thickness. The black lines do not necessarily delineate any particular fracture. They have been placed there to draw attention to the four directions of fractures, which are subparallel, not parallel. Figure 2, for comparison, shows fractures in an area where crystalline basement rocks are exposed. Most of these are macro-fractures, which are discussed in subsequent paragraphs. Figure 3 shows fractures in two other similar areas on the Canadian Shield. Figure 4a shows fractures or joints in the Wingate sandstone in east-central Utah; and Figure 4b, fractures, and dikes filling tension fractures, in basement rocks in Peru.

On aerial photographs, in sedimentary basin regions, two principal classes of fractures may be recognized: the macro-fractures and the micro-fractures. As to length, the one class grades into the other. An arbitrary unit of length for micro-fractures is $\frac{1}{2}$ mile. Few micro-fractures exceed five units or $2\frac{1}{2}$ miles in length. Macro-fractures, on the other hand, range from 2 to 50 miles, or even more, in length. Insofar as the sedimentary section is concerned, the macro-fractures are considered to have been generated at the level of basement. In contrast, the micro-fractures are considered to have been generated within the sediments themselves—some within the Palezoics, others much farther up in the section. It is considered probable, in fact, that the majority of the microfractures have their origin above the half-way point in the sedimentary column, regardless of its thickness. Even so, their mode or pattern of occurrence is strongly affected by deeper-seated structure.

To convey what is meant by *abundantly fractured*, it can be demonstrated that, normally, between 5 and 15 linear miles of fractures, in total, are present and discernible on high-resolution aerial photographs in each square mile of surface. This applies to plains regions in sedimentary basins. This figure, which in some areas may be double or triple this amount, excludes fractures of $\frac{1}{4}$ mile or less in apparent length.

Finally, to define the term, *systematically fractured*, we mean that there tends to exist a normal or regular fracture system in the earth's crust as a whole. This fracture system tends to be symmetrical with respect to the axis of rotation of the spinning earth and, consequently, with respect to the North and South poles and to the Equator. In accordance with the working hypothesis, this fracture system would approach complete regularity and symmetry if the crust were laterally homogeneous. It is considered that irregularity, and departures from symmetry, are the direct result of regional, or supra-regional, laterally heterogeneous conditions within the crust. More particularly, it is found that local departures from norm are due to local structure or to local stratigraphic anomalies. What we are looking for, in effect, is local disturbances in the otherwise regionally systematic, regular fracture pattern.

The four principal fracture directions, or *fracture sets*, of which the fracture system is composed, are arranged in two groups of two. In each group, the two

Fig. 2.—Exposed crystalline, basement rocks on Canadian Shield, northeast of sedimentary basin, showing strong and systematically arranged lineations.

Fig. 3.—Two other areas of exposed basement rocks, Canadian Shield, showing well developed, though complex, fracture system.

Fig. 4a.—Well developed fractures in east-central Utah.
b.—Fractures and dykes filling tension fractures in basement rocks in northwestern Peru.

61

sets are mutually at right angles or nearly so. One such conjugate pair is known as the *axial subsystem;* the other, as the *shear subsystem.* Figure 6 illustrates these two subsystems and how, together, they form the *fracture system* of an area, shown in the central part of the area. Normally, each fracture set is shown in a different color: red, purple, orange, and blue, representing the four sets, WNW, NNW, NNE, and ENE, respectively.

It is beyond the scope of this paper to delve into the theoretical mechanics

Fig. 5.—Vertical extension of fractures.

and mathematics of fracture generation and upward propagation. It must suffice to say that one of the principal causes of the systematic fracture system, so much in evidence to any painstaking investigator, is the rhythmic action of the tides in the crust of the earth itself. The amplitude of such earth tides, though only 9–14 inches, is, as a matter of fact, as much as two-thirds of the amplitude of the hydrospheric tides on the free ocean surface beyond the continental shelves, where their amplitude is only 18 inches. It is the cumulative effect, in shallow water, which gives the ocean tides their relatively great amplitude and considerable potential energy. In a similar manner, the earth tides gain their power from cumulative build-up. As a crustal stress, they are considered to be, by themselves acting alone, sufficient to have generated and to have caused the upward propagation of the fractures in the earth's crust, and, more particularly, in the sediments of large, deep, sedimentary basins. However, there are, as was previously stated, other groups of external stresses which have undoubtedly assisted in the process (Blanchet, 1957).

The fractures are probably generated by the process of fatigue, better known in metals which fail by rupture as the result of oft repeated stresses, such as long periods of vibration. Or it can be compared with the failure of a bridge by the cadence of marching soldiers. It is the cumulative effect of the small stresses set up by the terrestrial tides, applied repeatedly and rhythmically throughout millions of years (4 half-cycles per day, 1,460 times per year, $1\frac{1}{2}$ billion times per million years) that is one of the probable, principal causes of crustal fracturing of the kind discussed here.

To digress, for a moment, it can be safely supposed that if this tidal pulsation be reduced to the scale of a structure model, it would have to be represented by a fairly high-frequency vibration.

Incidentally, we have briefly considered only the principal lunar tide, M_2. In addition to this semidiurnal lunar tide, there are: the lunar diurnal tides, M_1; the solar diurnal and semidiurnal tides, S_1 and S_2; and three others of some significance, each with their own periods and wave lengths. In a lighter vein, it can be said that in some places and at certain times, veritable earth tide rips undoubtedly occur.

The propagation of fractures upward to the surface, even through the unconsolidated soils and glacial overburden, is considered to take place as a result of the flexing action of the tides within the layers of sediments, magnified by the lever-action or rocking motion of the larger basement blocks, and including, in particular, the isostatic blocks. The magnitude of this flexing action is indicated, for instance, by the variable flow of water from some deep springs. In eastern Africa, near the Great Rift Valley, for example, certain hot springs, in Katanga Province in the Belgian Congo, were being harnessed for hydro-electric power. It was discovered that the flow varied from 8.35 gallons per second to 16.1 and back to 8.35 in 12-hour cycles, in sympathy with the ocean tides (Vorster, 1956). Again, it is indicated by the extensive work of Japanese investigators who have found that the character and local magnitude of earth tides on either side of old faults, as measured with the delicate tiltmeter instruments, are distinctly different, even when the two measuring stations are only a few hundred yards apart (Nishimura, 1950).

Figure 5 shows the vertical extension of a fracture of the 150-foot wall of a small canyon in New York state. The depression shown at the top, can be followed in a straight line for some distance on the surface. There is a whole family of fractures here; they cause a minor offset in the creek.

To describe what is being accomplished by mathematically analyzing this fracture system and to present some idea of what a fracture-analysis survey consists, the ten steps involved are enumerated and outlined by use of a typical example, the Wizard Lake Upper Devonian biohermal reef mass in central Alberta.

1. The best obtainable aerial photography of the project area is procured. The scale of photography selected is governed by the degree of accuracy and amount of detail required in the end product. We have found that photography, taken with the new, R.C. 5 or R.C. 8 Swiss camera, to be ideal. This camera, which has a 6-inch focal length, essentially distortion-free lens, yields photography with the high resolution of fine detail essential for good fracture interpretation.

2. Template-controlled mosaics are prepared to ensure essentially true, relative orientation between all photographs. Uncontrolled mosaics for fracture analysis have a tendency to create false anomalies at each join between photographs.

3. Each photograph is interpreted for fractures. This is an exacting task because the great ma-

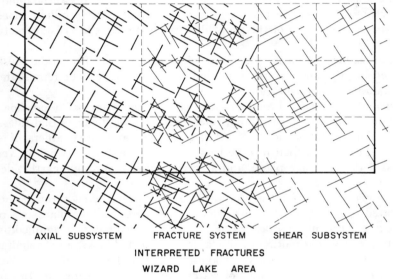

AXIAL SUBSYSTEM   FRACTURE SYSTEM   SHEAR SUBSYSTEM

INTERPRETED FRACTURES

WIZARD LAKE AREA

Fig. 6

jority of micro-fractures are by no means obvious. A well trained and experienced interpreter (a person with, among other qualifications, exceptionally good visual acuity) requires 5–7 hours to interpret each photograph. His interpretation is then checked by another interpreter and finally rechecked and edited by a supervisor.

4. The annotated fractures are then posted to transparent overlays on the controlled mosaics. Figure 6 shows a part of the fracture compilation in the Wizard Lake area. The small squares, outlined in broken lines, are square miles. The blank area, in which fractures are absent, in the southern row of sections, is Wizard Lake itself, just beyond the southern end of the Wizard Lake field. In this figure, only axial subsystem fractures are shown in the west part of the area, only shear subsystem fractures in the east part, and all fractures are shown in the central part of the area.

So long as the project area is essentially free of topographic relief, this posting of fractures is a simple, straightforward procedure. However, in areas where significant relief is present, all fractures which are not radial from the photo center have to be azimuth-corrected. This is done by using stereo-plotting instruments such as the Kail plotter.

5. The bearing of each fracture is determined; its length measured; and the co-ordinates of its center recorded.

6. The fractures are then analyzed to determine to which of the four sets each belongs.

7. Regional norms for each fracture set are then determined for the project area, or for several parts of the area, if large. This is accomplished by plotting a *fracture-azimuth frequency diagram*, similar to many such diagrams used in connection with faults, joints, and dikes. Figure 7 shows eight such frequency diagrams for eight areas more or less equally distributed throughout Alberta and western Saskatchewan. Each diagram represents a sample area of about 300 square miles. Each shows four sectors.

There is general consistency in direction of each sector throughout the whole region; for example, the azimuths of the regional norms for the NNE set: east to west across the top, 25°, 28°, and 30°; across the center, 24°, 26°, and 29°; and across the southern part, 25° and 27°. These gradual changes reflect two main factors. One is the gradual increase in azimuth of the NNE set with increasing latitude; that is, farther north. The other is the presence of regional structure.

8. The *statistical mean direction* of each fracture set is next calculated at each section corner, or closer, throughout the project area. This is accomplished by taking circular samples centered at each

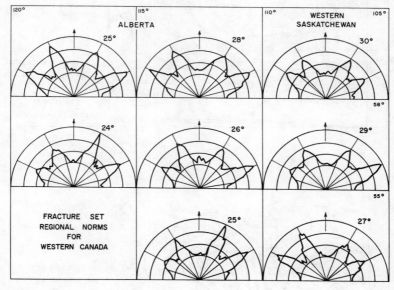

Fig. 7

section corner in turn. The diameter of a sample area is several miles. The sample diameter used depends on the depth of investigation required, the correct diameter to use being determined on the basis of empirical data. Appropriate weightings are applied, the greatest weight being given to fractures closest to the center point. This procedure yields *local mean directions* for each fracture set at each section corner. These values are analyzed and the *local deviations* are thereby obtained. These local deviations are then entered into empirical equations to yield what is known as the *structure intensity* of that point.

9. These structure intensity values are then contoured to yield the *structure intensity map* (Fig. 8), which is partly simplified for the present purpose. The area of the map is about one township. Below the map is a *structure intensity profile* along the line of section, CD. And below this, across the bottom of the figure, is the structure cross section, CD, showing the reef and its relative position with respect to the structure intensity data.

10. And lastly, this structure map is then structurally interpreted. As an aid to the structure interpretation, a *fracture-incidence map* is prepared in contour form. This map is based on a statistical count of the fractures present and shows the number of unit fractures per unit area. Further, maximum disturbance overlays are prepared and additionally, drainage anomaly studies are carried out to serve as additional confirmatory evidence of structure.

Figure 9 is the structure interpretation, shown by means of form lines and culmination axis, superimposed on the structure intensity data from Figure 8.

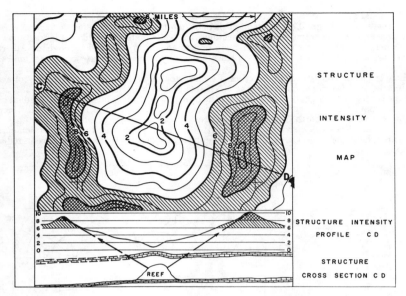

FIG. 8

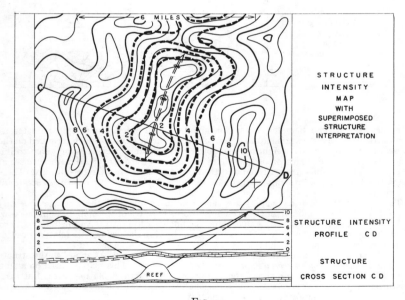

FIG. 9

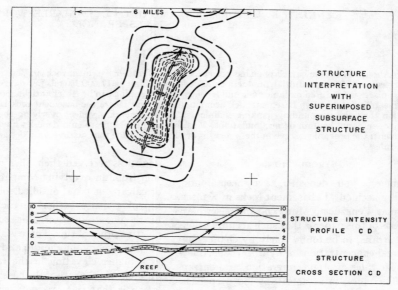

FIG. 10

Finally, Figure 10 shows a comparison between this interpretation and the bioherm as revealed by the drill. The bioherm is here shown structure-contoured in fine, broken lines. The structure interpretation of the intensity data is 2–2½ times the size of the anomaly it reflects from depth. This is usual, and is comparable with the areal expansion found in near-surface structures which are the result of supra-tenuous folding or draping by differential compaction over deep-seated structure.

That concludes the ten steps. The Wizard Lake biohermal reef mass was selected to illustrate the effectiveness of fracture analysis because this is one of our own test areas for developing empirical data. Because they would perhaps serve as better illustrations of the method and, at the same time, have been possibly more convincing, it would have been preferable to have shown and discussed reefs, domes, fault traps, and other structures that have been found *first* by fracture analysis and *second* by the drill. Unfortunately, such examples are shrouded in the cloaks of confidential security.

REFERENCES

BLANCHET, P. H., "Theoretical Basis of Fracture Analysis," *Manual of Photo Interpretation*, American Society of Photogrammetry, to be published during 1957.
NISHIMURA, E., 1950, "On Earth Tides," *Trans. Amer. Geophys. Union*, Vol. 31, No. 3 (June), p. 359.
VORSTER, W. H., 1956, "Generation of Electric Power from Hot Water Spring," *South African Mech. Engineer*, Vol. 5, No. 9 (April). Johannesburg, South Africa.

BULLETIN OF THE AMERICAN ASSOCIATION OF PETROLEUM GEOLOGISTS
VOL. 44, NO. 10 (OCTOBER, 1960), PP. 1682-1691, 4 FIGS.

PETROLEUM RESOURCES IN BASEMENT ROCKS[1]

KENNETH K. LANDES,[2] J. J. AMORUSO,[2] L. J. CHARLESWORTH, JR.,[2]
F. HEANY,[2] AND P. J. LESPERANCE[2]
Ann Arbor, Michigan

ABSTRACT

About 100 million barrels of oil has been produced so far from basement rocks in Venezuela, California, Kansas, and Morocco. Initial productions have been as high as 17,000 barrels per day. The reservoir in most places is fractured metamorphic and igneous rock. Trapping can be either anticlinal or due to varying permeability. All known basement rock accumulations occur where the basement rock is at a higher elevation than the surrounding flanking sediments. The sedimentary veneer overlying the basement rock may or may not contain oil accumulations. Most basement rock petroleum deposits have been found by accident; the probable reserve in these rocks is of such magnitude that discovery by design should become the rule.

INTRODUCTION

Commercial oil deposits have been found, largely by accident, in basement rocks in Kansas, California, Venezuela, Morocco, and elsewhere. The thesis of this paper is that such accumulations are not freaks, to be found solely by chance, but are normal concentrations of hydrocarbons, obeying all the rules of origin, migration, and entrapment. Therefore, in areas of not too deep basement, *oil deposits within the basement rocks should be sought with the same professional skill and zeal as accumulations in the overlying sediments.*

The four essentials for an oil pool, namely, source, reservoir rock, seal, and trap, are present in all of the basement rock fields so far discovered. Presumably there are infinitely more places yet to be discovered where these same conditions exist.

The only major difference between basement rock and overlying sedimentary rock oil deposits is that in the former case the original oil-yielding formation (source rock) can not underlie the reservoir. Basement rock accumulations obtain their oil from one of three possible sources: (1) overlying organic rock, from which the oil was expelled downward during compaction, (2) lateral, off-the-basement but topographically lower, organic rock from which oil was squeezed into an underlying carrier bed through which it migrated updip into the basement rock, and (3) lower lateral reservoirs from which earlier trapped oil was spilled due to tilting, or to overfilling. There is nothing radically different about any one of these sources. Many sedimentary rock reservoirs

doubtless received their oil by downward expulsion from an overlying compacting organic shale or calcareous ooze; oil-filled solution pores at the top of a dense carbonate rock section and beneath fine-grained organic rock furnish fairly positive proof of this. There is also ample evidence of spilled oil moving updip and becoming trapped at a higher elevation in the reservoir. The only difference in the case of the basement rock accumulation is that the reservoir changes type, from a permeable sedimentary rock to a permeable igneous or metamorphic rock, at the unconformity between the two. For example, where a sandstone laps upon, or abuts against, older fractured granite there will be no break in the continuity of the permeability, and the two rocks constitute a single reservoir.

In addition to the movement of oil into basement rock fractures by either downward expulsion or lateral migration there is also the possibility of it being sucked into the cracks by dilatancy during fracturing as described by McNaughton (1953) and referred to by Smith (1956, p. 384).

Most oil-field basement rocks are crystalline and brittle. Invariably and inevitably such rocks are fractured. This subject is treated with more detail in the following section. It is pointed out here, however, that the chances of drilling into a vertical or near vertical fissure in the basement rock fracture pattern are very slight, because the general rule in the past has been to penetrate the basement rock but a few feet at total depth. The advent of induced fracturing makes basement rock testing much more practicable than it was before.

The usual overseal ("cap rock") retaining basement rock oil accumulations is relatively tight

[1] Manuscript received, February 25, 1960.

[2] Members of Seminar in Advanced Petroleum Geology, University of Michigan, 1957.

1682

sedimentary rock plastered on the surface of the basement rock. However, a tight zone in the uppermost basement rock at the Mara field in Venezuela is barren and may act as a seal. At the other extreme is the situation in several California fields where a thick oil column extends from an oil-water interface within the basement rock upward through a continuous reservoir which includes such permeable material as wash and basal sandstone until a tight rock is reached somewhere above the base of the sedimentary section.

Basement rock oil traps are most likely to be convex upward anticlines or pseudo-anticlines. Anticlines are due to the local arching of the basement rock surface during a period (or periods) of diastrophism. Pseudo-anticlines (in terms of the basement rock surface) are brought about by the submergence and subsequent covering with sediment of hills sculptured in the basement rock during its emergence. Various modifications and inter-relations of these two basic trap types may exist. The vital points are that the reservoir is overlain by a convex upward seal, and that oil from superjacent or adjacent sedimentary layers could be squeezed or could migrate into this trap.

FRACTURE POROSITY IN BASEMENT ROCKS

Fracture porosity (Koester and Driver, 1953; Hubbert and Willis, 1955) is the void space between the walls of a crack or fissure. This space has very finite thickness, but the other two dimensions are indefinite. Except where cracks have been widened by solution the crack widths (void thicknesses) are probably in the one-tenth to one-fiftieth of an inch (0.25–0.05 cm.) range.

All rocks are brittle, to varying degrees, above the zone of flowage, so any movement of the earth's crust, from intense folding to gentle settling, will tend to fracture rock. Topography exerts some control on jointing near the surface according to Chapman (Chapman, 1958). There is also the fracturing produced by volume shrinkage; this in turn is brought about by cooling (igneous rocks) and desiccation (sedimentary rocks).

Fracturing tends to follow a geometric pattern. Although the shape of the pattern depends on the nature of the forces producing the fractures, all patterns have one feature in common: the fractures are interconnecting. The result is a network or system of voids with extreme permeability, providing that the crack widths are super-capillary.

The fracture porosity of a given volume of rock is the product of the average crack width and the area of the fracture planes. The latter is dependent on the pattern and its periodicity in space. Although fracture porosity, even at its best, falls far below the *maxima* reached by sandstone and carbonate rock reservoirs, it may still be adequate for the accumulation of hydrocarbons in commercial quantities. There is nothing unusual about this. Millions of people, scattered over the world, are dependent on fracture porosity of crystalline rocks (the basement rocks of the future) for their water supply.

The best place to study crystalline rock fracture porosity is at the outcrop and in water wells and mine workings. Most fractures are open to maximum width at the surface, due to solution and weathering. Walls that may be as much as several inches apart at the surface close to normal crack width within 200–300 feet. Near-vertical joints are the most common, with inclinations of 70°–90°. They are important water carriers and are generally continuous for considerable distances. Joint spacing is irregular and increases with depth.

Horizontal joints in general conform to the surface configuration of the rock. The spacing of this type is more regular than spacing of vertical joints, but the spacing increases with depth from about 1 foot apart near the surface to as much as 30 feet or more apart at a depth of 100 feet. It is probable that they do not exist as open fractures at depths greater than 200 feet.

Where crystalline rocks underlie the present surface adequate evidence exists of the presence of fractures, which are invariably water-filled below the water table. Some of these data have been obtained in underground mine workings where shafts have been sunk to exploit ore deposits. In this case the water moving through the fractures is a detriment for it has to be pumped from the mine. On the other hand, wells are dug or drilled in crystalline rock areas for local water supplies, virtually all of which comes from fractures.

Fuller (1906) tabulated water conditions in 9 "deep" mines in igneous and metamorphic rocks in the United States. The maximum depths ranged from 1,200 to 5,000 feet. Most of these mines reported that the greater part of the water was in the first 600–1,000 feet below the surface, but one mine which reached a total depth of 1,500 feet reported "joints everywhere saturated with water, no diminution with depth."

Water-well data in crystalline rocks were assembled by Ellis (1909) for Connecticut and by Clapp (1909) for Maine. Data tabulated for 25 wells show a range in depth to the "principal water supply" from 60 to 800 feet, with an average of 309 feet. Most wells stopped drilling a few feet or a few tens of feet below the top of the "principal source"; one well, however, went to 1,465 feet without finding a yield greater than 3 gallons per minute (103 barrels/day). The average water yield of these wells, converted to barrels per day, was 1,166 with a maximum of 6,860 barrels.

The discrepancy between the water depths in crystalline rock mines and in wells is due to the fact that the mines report the bottom of the principal water body and the wells report the top.

It can be concluded that water-bearing fractures are a general rule in crystalline rock terrane between the water table and depths below the surface of 600–1,000 feet.

If the present crystalline rock surfaces were to become basement surfaces by submergence and deposition of overlying sediments, even to a thickness of thousands of feet, the present 600–1,000-foot vertical section of fractures would not be obliterated short of penetrating the zone of flowage. Irregularities in the smoothness of the fissure walls, including outward convexity below the original outcrop, would prevent the cracks from being squeezed tightly shut; also the spalling of rock fragments off the fracture walls and the falling and washing in of foreign material would prop the cracks open. Lastly the presence of fluids under pressure in the fractures would be a further deterrent to squeezing shut.

The basement rocks of today were similarly exposed in the geological past and must have had a similar history.

The best evidence that fracture systems of considerable extent can and do exist in basement rock even where buried beneath thousands of feet of sediment is the fact that water in large volume and oil in commercial quantities have been found stored in reservoirs of this type.

PRESENT BASEMENT ROCK PRODUCTION

Commercial production of oil from crystalline basement rocks is not wishful thinking but an established fact. Already, in round numbers, 100 million barrels of oil have been produced from such reservoirs, mainly from four fields in Venezuela and California. Initial productions range as high as 17,000 barrels per day. In most of these fields the basement rock is but one of several reservoirs; in several of the nineteen fields reported as having basement rock production only one well is producing from that reservoir.

Venezuela.—The La Paz and Mara oil fields of western Venezuela are outstanding exceptions to the generalization that basement rock production is found by accident (Smith, 1956). In the La Paz field it was observed that the oil being produced from Cretaceous limestone was stored principally in fractures which are most abundant on the anticlinal crest, and it was reasoned that if the underlying upfolded and upfaulted basement rocks were likewise fissured they too might contain oil. Furthermore, material balance studies "suggested that the volume of fluids expanding in the reservoir was larger than could be contained in the calculated pore volume in the limestone" (page 381). The first two wells purposely drilled into the basement at La Paz were not successful, but the third well, the Shell Company's P-86, was completed in April, 1953, at total depth of 8,889 feet, which is 1,089 feet below the top of the basement complex, for an initial production of 3,900 barrels per day of 34° oil. As the limestone reservoir had been cased off, this was solely basement rock oil. Since then twelve other wells have been drilled at La Paz with an average penetration of the basement of 1,650 feet and a maximum of 3,087 feet. The average initial production has been 3,600 barrels per day, but one well had an initial yield of 11,500 barrels.

The basement reservoir has also been successfully developed in the adjacent Mara field where 29 wells have been drilled to the basement for an average initial production of 2,700 barrels per day. One of these wells found the overlying Cretaceous reservoir tight and completely non-productive, but came in for 17,000 barrels per day initially from the basement rock. During July, 1955, the daily production rate in these two fields was 76,700 barrels, and the cumulative production of oil from the basement was 31 million barrels.

The structural and stratigraphic relations in the La Paz field district are shown by cross section (Fig. 1). The basement rocks are a complex of metamorphic and igneous rocks with the metamorphics most abundant along the axis of the faulted anticline. Although the oil in both the Cretaceous and basement reservoirs is practically identical, bottom-hole-pressure measurements in the Mara field indicate separate reservoirs. Possible seals separating the two reservoirs are an unproductive and apparently tight zone at the top of the basement, an overlying well-cemented sandstone, or a marl and shale section which occurs about 250 feet above the basement.

Smith (page 385) believes that the La Luna formation which overlies the Cretaceous limestone reservoir is the probable source of the oil for both the limestone and basement rock reservoirs.

California.—California has at least five fields which obtain oil from basement rocks. Relatively few wells produce from the basement rocks alone; most are multiple completions in both the basement schist and the overlying schist conglomerate. Descriptions of basement oil occurrence in each of the five fields follow.

The Playa Del Rey (Barton, 1931; Metzner, 1935, 1943) oil field lies along the ocean shore at Santa Monica Bay in Los Angeles County. It is subdivided into two areas: (1) Ocean Front or Venice, which produces from a Pliocene reservoir as well as from basal sediments and basement; and (2) Del Rey Hills which is productive from the deep zone only. The Playa Del Rey structure consists of a northwest-southeast-trending ridge of schist of possible Jurassic age. This ridge was exposed to erosion, and drainage channels were cut into its flanks, until submergence in late Miocene time. A basal conglomerate of angular schist fragments and a quartz sand were deposited as aprons around the ridge, with the greatest thickness in the submerged drainage channels. Deposition of this basal formation ended while the ridge summit was still above sea-level. The basal conglomerate and sand are not continuous between the Venice and the Playa Del Rey Hills areas, the two being separated by the main schist ridge. The conglomerate ranges from 0 to 234 feet in thickness. A dark brown compact shale, known as the "nodular shale" because it contains calcium phosphate nodules, conformably overlies the basal conglomerate. It ranges between 60 and 308 feet in thickness throughout the field. Both it and the underlying conglomerate are buttress formations. Further deposition during Miocene time laid 500 feet of black shale and sandy shale conformably on the nodular shale. A lower Pliocene series of shales and oil sands 1,000–1,100 feet thick was deposited on the Miocene deposits. This is the upper productive zone. Pliocene and Pleistocene deposits of shales, shales with sand lenses, and sandy shales overlie the Miocene sediments with about 150 feet of Pleistocene sand and gravel at the top.

The anticlinal trap of the Pliocene sediments was evidently primarily due to the settling of sediments over the schist ridge rather than to

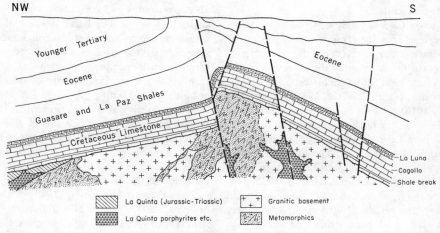

Fig. 1.—Cross section (after Smith) of LaPaz field, western Venezuela. Scale approximately 1/80,000.

diastrophic movement. Depth to the ridge ranges from 5,300 to 6,800 feet. The deeper productive zone, found in both areas, consists of a basal conglomerate overlying the schist and underlying the nodular shale. Although this conglomerate is the main oil reservoir in the lower zone, fractures in the nodular shale and the schist are also productive. No production is obtained exclusively from the schist.

The El Segundo (Porter, 1938; Reese, 1943) oil field which is about 4 miles south of Playa Del Rey underlies the western part of the town of El Segundo. The field is divided into two areas by a northwest-trending zone of faulting. The eastern part produces from a basal conglomerate made up of schist pebbles, and the western part produces directly from fractures in the schist itself. Discovery of the field took place on August 24, 1935, when the El Segundo No. 1, which was drilled east of the fault zone, came in for about 400 barrels of 28° gravity oil a day. The western part of the field began producing on May 23, 1937, when the Security No. 1 flowed 4,563 barrels of 26.5° gravity oil from fractured schist at a depth of 7,253 feet. This discovery was accidental, for the well was probing for schist conglomerate. Sixty-six wells were drilled in the development of this field which covers 950 acres. Wide variations in production between adjacent wells are characteristic. As at Playa Del Rey, Miocene sediments laid down by a transgressing sea drape over a schist hill (Fig. 2) festooned with schist conglomerate. Here, too, nodular shale directly overlies the conglomerate, or the schist where the conglomerate is absent.

El Segundo field has only one producing zone. In the eastern area the reservoir transgresses the conglomerate-schist contact; on the western side conglomerate is absent and the reservoir is the fractured basement schist exclusively.

The Santa Maria (Canfield, 1943; Cabeen and Sullwold, 1946) oil field is in the central part of Santa Maria Valley in the northwestern corner of Santa Barbara County. Discovery was on July 15, 1934, with a 50-barrel well. The field did not assume importance, however, until a 2,376-barrel-a-day well was completed on March 31, 1936. Production was from various Miocene zones, in particular, fractured Monterey chert. Three hundred wells had been drilled on more than 6,000 acres when in December, 1942, W. R. Gerard's Acquistapace No. 1 flowed 2,500 barrels a day of 18.5° gravity oil from the fractured

Knoxville sandstone basement. This discovery was accidental, brought about by lack of knowledge of the elevation of the basement. Up to June, 1946, more than 40 wells proving 420 acres of Knoxville had been completed with some of the wells more than 1,200 feet deep in the Knoxville. Most of the wells pumped 200–400 barrels of oil a day. The Knoxville supplies the only production in the field north of the Monterey limits.

There are two unusual features about the basement oil accumulation at Santa Maria. First, the reservoir is a fractured sandstone instead of crystalline rock. Second, the trap is not anticlinal, but is due to a lithologic change of the basement rock from fractured sandstone to presumably unfractured shale up the flank of the Santa Maria syncline, which is on the north. The trapping in the overlying Miocene sediments is also stratigraphic due to updip pinch-out of the reservoirs, all of which are of the fracture type.

The Wilmington (Brown and Kew, 1932; Winterburn, 1943; Cabeen and Sullwold, 1946) oil field is in the city of Wilmington and includes the harbor area of Long Beach. The discovery well was completed on January 26, 1932. Schist production was established in May, 1945, by the Union Pacific E-47 well which unexpectedly encountered schist at minus 5,787 feet. It flowed 387 barrels a day. In 1946 rates of 1,200–2,000 barrels a day were established for wells producing from the schist reservoir. Wilmington has produced more than 22 million barrels of oil from the basement rocks.

Jurassic basement schist hills are unconformably overlain by 3,200–3,800 feet of Miocene shales and sands. The basal 100 feet of the Miocene section consists of a schist conglomerate. The structure of the field is a northwest-southeast-trending anticline plunging northwest. It is cut by transverse normal faults some of which branch out into smaller faults.

Six productive zones are present in the Wilmington field. Of these, five are in Pliocene and Miocene sediments and the sixth is the basement rock fractured schist. The productive zone in the schist does not extend below minus 6,200 feet, the elevation of the oil-water interface.

The Edison (Edwards, 1943; Cabeen and Sullwold, 1946; National Oil Scouts, 1956) oil field is approximately 9 miles southeast of Bakersfield, Kern County, on the east side of the San Joaquin Valley. This field was 11 years old when the basement reservoir was discovered in July,

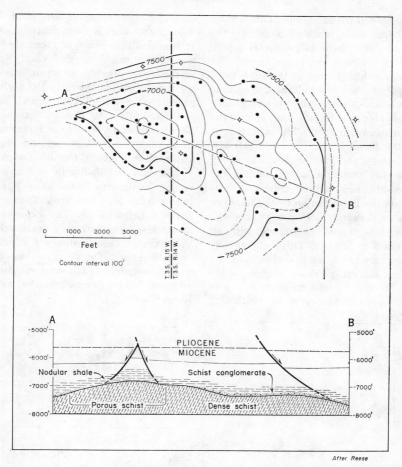

After Reese

Fig. 2.—El Segundo oil field, California (after Reese).
Upper: Contours on top of schist basement rock. All contour figures are in feet below sea-level.
Lower: Generalized section along line AB. Oil reservoirs are porous schist on west and schist conglomerate on east. Depths in feet below sea-level.

1945, by a well which accidentally drilled 100 feet into the schist and was completed for 528 barrels (initial daily production) of 19.2° gravity oil. Some later basement wells had initial potentials of more than 1,000 barrels per day. A total of 805 wells had been drilled in the field by the end of 1955, of which 114 were drilled into the schist and 111 were still producing from that reservoir. The total production during 1955 was 4,950,962 barrels, of which 434,697 barrels (9%) came from the schist. The cumulative production from the fractured schist is in excess of 20 million barrels.

The basement schist produces from hard fractured rock below the weathered zone, with accumulation probably the result of migration updip from one of the buttressing sedimentary formations. No relation between production and depth drilled into the schist has been noted.

Kansas (Landes, 1939; Walters, 1946, 1953; Walters and Price, 1948).—All of the oil fields in this state which currently produce from Precambrian basement rocks lie on the Central Kansas uplift in Russell, Ellsworth, Barton, and Rice counties (Fig. 3). The Precambrian rocks are quartzite, schist, gneiss, and granite. Clear granular quartzite is the rock most commonly penetrated; it occurs on the summits of many buried Precambrian hills. Muscovite and biotite schists and feldspar-biotite gneisses are found down the flanks of the quartzite hills. Coarse-grained, pink granite is also common and it too occurs in buried hills. The basement rock is overlain by Cambro-

Ordovician sediments which have been divided by Walters into four rock types or members: (1) basal sandstone, (2) sandy dolomite, (3) coarse white dolomite, and (4) cherty dolomite. The basal sandstone, which ranges in thickness from 0 to 80 feet, is a porous, poorly indurated sandstone consisting of coarse, subangular quartz grains. Arkosic material is present at the contact with the underlying Precambrian rocks. The sandstone is confined to the broad lowland areas which surround the quartzite and granite hills. The sandy dolomite member unconformably overlies the basal sandstone in the broad low areas and overlaps directly on the Precambrian on the flanks of each hill. Above the sandy dolomite is the white, coarsely crystalline dolomite member, with an average thickness of 120 feet. This member may be present not only in the areas surrounding the buried hills but also on the crests of some of the buried hills. It contains numerous caverns and sinkholes where it was subjected to weathering

during early Pennsylvanian time. The cherty dolomite member is approximately 300 feet thick. It contains thin zones of green shales. Like the underlying member, the cherty dolomite is leached considerably where it was exposed to weathering during the early Pennsylvanian. It is present between the Precambrian hills only.

Tectonically, the Central Kansas uplift is a broad arch which trends northwestward. It has a Precambrian crystalline core and is flanked by upturned and beveled Cambro-Ordovician, Ordovician, and Mississippian sediments. Pennsylvanian sediments both abut against buried Precambrian hills and are draped over them in gentle anticlinal folds. The Pennsylvanian sediments also unconformably overlie the upturned and beveled older sediments which flank the crystalline core. The cross section (Fig. 4) illustrates the structural and stratigraphic relations of a typical buried Precambrian hill in the Central Kansas uplift.

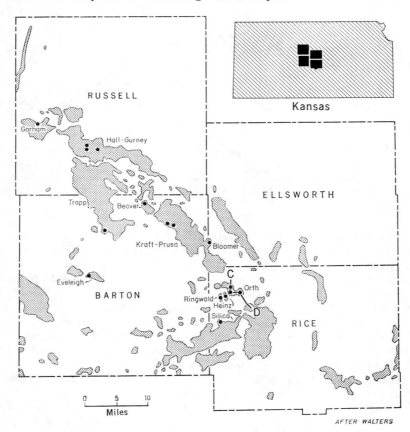

Fig. 3.—Map showing central Kansas oil fields, and oil wells producing from fractured Precambrian basement rocks. CD is line of section shown on Figure 4. After Walters (1953).

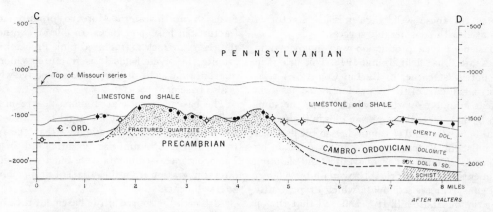

Fig. 4.—Section along line CD on Figure 3. Only the wells penetrating pre-Pennsylvanian rocks are shown. Vertical exaggeration ×10. Depths in feet below sea-level. After Walters (1953).

The buried Precambrian surface in the area under discussion is a relatively smooth, gently arched surface on which are hills which may rise 700 feet above the general level. The slopes of the hills may be as much as 10°, but are generally less. The Arbuckle dolomite has a regional northwest-southeast strike, parallel with the general trend of the uplift, and dips $\frac{1}{2}$°–1° northeastward and southwestward away from the uplift axis. The Arbuckle beds may have initial dips, however, of 2°–10° on the flanks of the Precambrian hills.

According to Walters, at the beginning of Arbuckle time the Precambrian rock surface was a monadnock-studded peneplain which had developed across a terrane of igneous and metamorphic rocks. The monadnocks were mainly of resistant quartzite. In early Cambro-Ordovician time the hills were slowly encroached upon by transgressing seas, and by late Cambro-Ordovician time the hills were buried beneath about 340 feet of dolomite. During the Ordovician the Simpson and other formations were deposited. At the close of the Devonian a broad arch rose and was eroded so that Arbuckle dolomite was exposed at the core of the uplift. That arch trended in a more northwesterly direction than the present-day Central Kansas uplift. Superimposed on it were three parallel anticlines. During the Mississippian period the seas transgressed onto the arch and perhaps covered it. Prior to Pennsylvanian deposition arching took place in a more northerly direction than the older arch, and in a narrower zone, producing the present Central Kansas uplift. However, the older structure is represented by the three parallel anticlinal axes which cross the uplift northwest and southeast.

The pre-Pennsylvanian arching was accompanied and followed by erosion of older rocks down to the Precambrian core of the uplift. During this erosion hills in the Precambrian were resurrected, rising above the newly peneplaned surface and surrounded by solution valleys in the limestone terrane. These valleys became filled with "Pennsylvanian conglomerate." In late Des Moines time the hills again became islands in the transgressing Pennsylvanian sea. By the end of the Missouri epoch the hills were covered by 250 feet of sediments and throughout Virgil time deposition continued until the region was buried under about 1,000 feet of Pennsylvanian sediments. Later, more than 1,800 feet of Permian sediments were deposited and these in turn were partly eroded and subsequently covered by Cretaceous sediments.

Figure 3 shows the location of the fields which produce from basement rocks. Precambrian quartzite production was discovered in 1933 in the Orth field, in northwestern Rice County. The oil occurs in fractured Precambrian quartzites on the summits of the buried hills (Fig. 4). The quartzite itself is impervious. About 1,243,000 barrels of oil had been produced from 16 quartzite wells in the Orth field to January 1, 1952.

The Ringwald field, which is about $1\frac{1}{2}$ miles southwest of the Orth field, was discovered in 1949. The structure and stratigraphy here are similar to the Orth field. There is a Precambrian quartzite hill with Lansing-Kansas City (Pennsylvanian) limestones abutting and draped over it in an anticlinal fold. The field contains six wells which are producing or have produced from fractured Precambrian quartzites. During the devel-

KENNETH K. LANDES, *ET AL.*

opment of these wells high-pressure fracturing was used with considerable success.

Basement rock production was discovered in the Kraft-Prusa field around 1948. This field is in the northeast corner of Barton County. A relatively small part of the oil production comes from a clear granular quartzite which forms the summits of three large Precambrian hills in the area. A basal Paleozoic sand (Reagan) lies on the flanks of the hills; the Reagan is conformably overlain by Arbuckle dolomite: The quartzites and flanking sediments are unconformably overlain by 900 feet of Pennsylvanian sediments which drape gently over the quartzite hill. It is believed that the oil migrated from the overlying Pennsylvanian rocks into fractures in the quartzite.

The Beaver, Bloomer, Trapp, Eveleigh, and Silica fields are similar to those previously mentioned in that there is some production from fractured Precambrian quartzites occurring in buried hills.

The Hall-Gurney and Gorham fields are similar to each other; each has several wells which produce from a fresh pink biotite granite. Drilling has shown that there is a very erratic distribution of oil in the granite. One well encountered porosity and oil 70 feet below the top of the granite, while many others have failed to find any oil.

All of the areas of basement rock production as described by Walters have the following important factors in common.

1. Every known quartzite reservoir occurs in the upper part of a quartzite hill which rises from 100 to 500 feet above the surrounding basement rock. No oil has been found in structurally low areas.

2. On the flanks of each hill are oil-bearing Cambro-Ordovician Arbuckle beds which are as much as 500 feet in thickness.

3. Pennsylvanian beds rest directly on the Precambrian in every area where there is basement rock production.

4. The oil in the fractured Precambrian quartzites occurs beneath the pre-Pennsylvanian unconformity.

5. The oil is believed to have migrated into the quartzite from either the flanking Cambro-Ordovician beds or from the overlying Pennsylvanian rocks.

Morocco (Bruderer *et al.*, 1951; Bruderer, 1952; La Soc. Cherifienne, 1951; Burger, 1957, 1958; Levy, 1956; Moyal, 1956).—At least eight oil fields in northwestern Morocco produce from fractures in basement rocks. In one or two fields the reservoir rock is fractured pink Precambrian granite; in seven fields it is fractured chloritic quartzite and shale or slate of Paleozoic age which is also basement rock.

The basement rock accumulations are in upfaulted blocks which are marginal to a Mesozoic sedimentary basin. It is assumed that the fractures in the Precambrian granite and in the Paleozoic meta-sediments were filled with oil by lateral migration from the adjacent deeper Mesozoic sediments.

Although the source of the basement rock oil is probably the sediments in the adjacent Mesozoic basin so far, at least, more oil has been produced from the basement rocks than from the sediments. By the end of 1957 the basement fracture reservoirs had produced more than $3\frac{3}{4}$ million barrels of oil since discovery in 1947.

Other areas.—There are other fields which produce in part at least from basement rock. In the Texas Panhandle the Granite Wash reservoir, which gets all of the credit in the production statistics, appears to continue downward into fractured granite reservoir in some of the fields [3]

CONCLUSIONS

From this review of known commercial oil occurrences in basement rocks the following conclusions appear justified.

1. Oil can and does occur in basement rocks in profitable quantities in at least twenty-five oil fields on three continents. Therefore, the potential oil reservoir section includes not only the sedimentary veneer but also the upper part of the basement rock complex.

2. Basement rock accumulations are found both below oil pools in the sedimentary rock section, and where no superjacent oil deposits are known.

3. The reservoir may or may not be continuous across the sedimentary rock-basement rock contact. Where the reservoir is confined to the basement rock the seal may be either the lowermost sedimentary rock or, more rarely, the uppermost basement rock.

4. Trapping within the basement is not exclusively anticlinal; varying permeability (as at Santa Maria, California) may also trap the oil.

[3] William J. Vaughn, Jr., oral communication, March 18, 1960.

5. No well entering the basement at a relatively high elevation and not finding water therein should be abandoned until it has penetrated several hundred feet and has been given a fracture treatment in order to break into the basement rock fracture system.

6. Structurally high producing fields in areas where the basement surface is within feasible drilling depth should be explored downward through the sedimentary section and into the basement.

7. In suitable areas exploratory wells should be located deliberately in order to probe upward bulges of the basement rock as well as the sedimentary veneer. Contour maps of the basement rock surface, such as those already published for Texas and southeast New Mexico (Flawn, 1956), the Texas and Oklahoma panhandles (Totten, 1956), northeastern Oklahoma (Ireland, 1955), Kansas (Farquhar, 1957), South Dakota (Petsch, 1953), and the west side of the Los Angeles basin (Schoellhamer and Woodford, 1951), can be of great value in planning explorations of this type.

REFERENCES

BARTON, C. L., 1931, "A Report on Playa Del Rey Oil Field, Los Angeles, County, California," *California Oil Fields*, Vol. 17, No. 2 (October-December), pp. 5–15.

BROWN, A. B., AND KEW W. S. W., 1932, "Occurrence of Oil in Metamorphic Rocks of San Gabriel Mountains, Los Angeles County, California," *Bull. Amer. Assoc. Petrol. Geol.*, Vol. 16 (August), pp. 777–85.

BRUDERER, W., 1952, "Morocco (French)," *ibid.*, Vol. 36 (July), pp. 1413–15.

———, et al., 1951, "French Morocco," *ibid.*, Vol. 35 (July), pp. 1610–12.

BURGER, J. J., 1957, "Morocco," *ibid.*, Vol. 41 (July), pp. 1575–79.

———, 1958, "Morocco," *ibid.*, Vol. 42 (July), pp. 1667–70.

CABEEN, W. R., AND SULLWOLD, H. H., JR., 1946, California Basement Production Possibilities," *Oil Weekly* (June 22), pp. 17–25.

CANFIELD, C. R., 1943, "Santa Maria Valley Oil Field," *California Dept. Nat. Res. Bull. 118*, pp. 440–42.

CHAPMAN, CARLETON A., 1958, "Control of Jointing by Topography," *Jour. Geol.*, Vol. 66 (September), pp. 552–58.

CLAPP, F. G., 1909, "Underground Waters of Southern Maine," *U. S. Geol. Survey Water-Supply Paper 223*.

EDWARDS, E. C., 1943, "Edison Oil Field," *California Dept. Nat. Res. Bull. 118*, pp. 576–78.

ELLIS, E. E., 1909, "Ground Water in Crystalline Rocks in Connecticut," *U. S. Geol. Survey Water-Supply Paper 232*, pp. 54–103.

FARQUHAR, O. C., 1957, "The Precambrian Rocks of Kansas," *Kansas Geol. Survey Bull. 127*, Pt. 3.

FLAWN, PETER T., 1956, "Basement Rocks of Texas and Southeast New Mexico," *Univ. Texas Pub. 5605*.

FULLER, M. L., 1906, "Total Amount of Free Water in the Earth's Crust," *U. S. Geol. Survey Water-Supply Paper 160*, pp. 59–72.

HUBBERT, M. KING, AND WILLIS, DAVID G., 1955, "Important Fractured Reservoirs in the United States," *Proc. 4th World Petrol. Congress*, Sec. I, pp. 57–82.

IRELAND, H. A., 1955, "Precambrian Surface in Northeastern Oklahoma and Parts of Adjacent States," *Bull. Amer. Assoc. Petrol. Geol.*, Vol. 39, pp. 468–83.

KOESTER, EDWARD A., AND DRIVER, HERSCHEL L. (chairmen), 1953, "Symposium on Fractured Reservoirs," *ibid.*, Vol. 37 (February), pp. 201–330.

LANDES, KENNETH K., 1939, "Oil Production from Precambrian Rocks in Kansas" (abst.), *Oil Weekly* (March 27), p. 74.

LA SOCIETE CHERIFIENNE DES PETROLES, 1951, "Un Gisement dans les Schistes Fractures du Socle l'Oued Beth (Maroc Francais)," *Proc. 3d World Petrol. Congress*, pp. 315–28.

LEVY, R., 1956, "Morocco," *Bull. Amer. Assoc. Petrol. Geol.*, Vol. 40 (July), pp. 1612–20.

McNAUGHTON, D. A., 1953, "Dilatancy in Migration and Accumulation of Oil in Metamorphic Rocks," *ibid.*, Vol. 37 (February).

METZNER, L. H., 1935, "The Del Rey Hills Area of the Playa Del Rey Oil Field," *California Oil Fields*, Vol. 21, No. 2 (October-December), pp. 5–26.

———, 1943, "Playa Del Rey Oil Field," *California Dept. Nat. Res. Bull. 118*, pp. 292–94.

MOYAL, MAURICE, 1956, "SCP Finds 5 New Oil Areas in Morocco During 1955," *World Oil* (July), pp. 176 *et seq.*

NATIONAL OIL SCOUTS AND LANDMEN'S ASSOCIATION, 1956, *Yearbook*.

PETSCH, B. C., 1953, "Map of Precambrian Surface, South Dakota," *South Dakota Geol. Survey*.

PORTER, L. E., 1938, "El Segundo Oil Field," *Trans. Petrol. Dev. A.I.M.E.*, Vol. 127, pp. 81–91.

REESE, R. G., 1943, "El Segundo Oil Field," *California Dept. Nat. Res. Bull. 118*, pp. 295–96.

SCHOELLHAMER, J. E., AND WOODFORD, A. O., 1951, "The Floor of the Los Angeles Basin . . . ," *U. S. Geol. Survey Map OM 117*, Sheet 1, Oil and Gas Inv. Ser.

SMITH, J. E., 1956, "Basement Reservoir of La Paz-Mara Oil Fields, Western Venezuela," *Bull. Amer. Assoc. Petrol. Geol.*, Vol. 40 (February), pp. 380–85.

TOTTEN, ROBERT B., 1956, "General Geology and Historical Development, Texas and Oklahoma Panhandles," *ibid.*, Vol. 40, pp. 1945–67.

WALTERS, ROBERT F., 1946, "Buried Pre-Cambrian Hills in Northeastern Barton County, Central Kansas," *ibid.*, Vol. 30, No. 5 (May), pp. 660–710.

———, 1953, "Oil Production from Fractured Pre-Cambrian Basement Rocks in Central Kansas," *ibid.*, Vol. 37, No. 2 (February), pp. 300–13.

———, AND PRICE, ARTHUR S., 1948, "Kraft-Prusa Oil Field, Barton County, Kansas," *Structure of Typical American Oil Fields*, Vol. 3, Amer. Assoc. Petrol. Geol., pp. 249–80.

WINTERBURN, R., 1943, "Wilmington Oil Field," *California Dept. Nat. Res. Bull. 118*, pp. 301–05.

THE BULLETIN OF THE AMERICAN ASSOCIATION OF PETROLEUM GEOLOGISTS
Vol. 45, No. 1, January, 1961

REGIONAL STUDY OF JOINTING IN COMB RIDGE-NAVAJO MOUNTAIN AREA, ARIZONA AND UTAH[1]

ROBERT A. HODGSON[2]
Pittsburgh, Pennsylvania

ABSTRACT

The spatial relations of joints and, in particular, structural details of individual joints, offer clues of their origin. Important features of joints have been largely neglected in previous joint studies and the present study is an attempt to determine more closely the true nature of joints in sedimentary rocks and to suggest a mode of origin more in line with field observations than is present theory.

The study area comprises about 2,000 square miles of the Colorado Plateau in northeastern Arizona and southeastern Utah where sedimentary rocks ranging from Pennsylvanian to late Cretaceous in age are exposed.

A simple, non-genetic joint classification is presented based on the spatial relations of joints and the plumose structures on joint faces. Joints are grouped as systematic or non-systematic with cross-joints defined as an important variety of non-systematic joints.

Plumose structures on joint faces indicate that joints are initiated at some structural inhomogeneity within the rock and propagated outward, thus precluding movement in the direction of the joint faces at the time of formation. Spatial relations of systematic joints point to formation at or near the earth's surface in a remarkably homogeneous stress field.

The regional joint pattern is composed of a complex series of overlapping joint trends. The pattern as a whole extends through the entire exposed rock sequence. Intersecting joint trends have no visible effect on each other and may terminate independently in any direction. Each joint trend of the regional pattern crosses several folds of considerable magnitude but does not swing to keep a set angular relation to a fold axis as the axis changes direction.

Hypotheses stating that joints are related genetically to folding are rejected for the mapped area. The shear, tension, or torsion theories of jointing require that only one or two sets of joints be considered as the result of a particular stress condition. Where more sets are present, different stress conditions must be postulated for each set or pair of sets believed to be related genetically. The joint pattern in the mapped area can not be interpreted in such terms without making these assumptions. Alternatively, in accord with theoretical and experimental evidence, semi-diurnal earth tides are considered as a force capable of producing joints in rocks through a fatigue mechanism. Field observations suggest that joints form early in the history of a sediment and are produced successively in each new layer of rock as soon as it is capable of fracture. The joint pattern in pre-existing rocks may be reflected upward into new, non-jointed rock and so control the joint directions.

Much critical data from areas with different geologic histories are needed before a quantitative evaluation of this hypothesis can be made. The question of the ultimate origin of the regional joint pattern and its genetic relation, if any, to other structure at depth can not be answered on data now available.

INTRODUCTION

Joints are one of the most common structural features of the earth's crust, yet they have been given little more than passing notice by most geologists. Their mode of origin is highly contro- versial and diametrically opposed opinions con-

[1] Manuscript received, April 22, 1960. This paper is a part of a dissertation presented to the Faculty of the Graduate School of Yale University in candidacy for the degree of Doctor of Philosophy.

[2] Gulf Research and Development Company. The writer expresses deepest appreciation to John Rodgers and to members of the geological staff of the Carter Research Laboratory who gave generously of their time and experience in frank and helpful criticism through the course of the investigation. He also acknowledges gratefully the financial support given by the Department of Geology, Yale University, through the Donnel Forster Hewitt Fund, summer, 1955, and by the Carter Research Laboratory, summer, 1956.

1

cerning their origin are not uncommon. The chart, Figure 26, shows, in graphic form, several of the major hypotheses advanced to account for joints and the sequence in which they appear in the literature.

Present theories of the origin of joints in sedimentary rocks are based primarily on the observation and interpretation of their more obvious features such as parallelism, the angular relations in plan between joint sets and between joint sets and other structural features such as folds and faults. There are, however, additional features of joints that are seldom discussed and which throw light on their mode of genesis.

The present work is restricted to a study of joints in the sedimentary rocks cropping out in the Comb Ridge-Navajo Mountain area, Arizona and Utah. The objectives of this study are first, to record the physical features of individual joints and second, to define the regional joint pattern and its areal relations to the folds of the area. The ultimate objective is to use these data to suggest the origin of the joints.

The Comb Ridge-Navajo Mountain area was selected for study for three important reasons: (1) joints are readily observed in plan and section through a relatively thick sedimentary section comprising a variety of rock types; (2) other structural features of the area are simple and already separate so that the relations of the regional joint pattern to these features can be observed directly; and (3) the dry climate minimizes the effects of weathering that commonly destroy or modify significant features of joints.

GENERAL FEATURES

The area mapped comprises about 2,000 square miles of the Colorado Plateau in northeast Arizona and southeast Utah. It is bounded on the north by the San Juan River and the parallel of 37° 15′ North Latitude and on the south by the parallel 36° 15′ North Latitude. The eastern and western limits of the region are the meridians of 109° 30′ and 111° 00′ West Longitude, respectively (Fig. 1). Almost the entire area lies within the boundaries of the Navajo Indian Reservation.

The Comb Ridge-Navajo Mountain area lies in the central part of the Colorado Plateau physiographic province and comprises parts of the Monument Valley and Navajo Uplands subprovinces (Gregory, 1917; Kiersch, 1955). The average elevation is about 6,000 feet.

The Monument Valley subprovince lies in the north-central part of the mapped area and is an upland intricately dissected by canyons and dry washes lying between colorful and unusual buttes and mesas. This subprovince is bordered on the east and south by the Comb Ridge monocline and on the west by Organ Rock anticline (Fig. 4).

The Navajo Uplands subprovince comprises a high structural plain that rims the Monument Valley. Much of the surface of the Navajo Upland is underlain by Navajo sandstone and covered by loose or semi-stabilized deposits of wind-blown sand.

Throughout the mapped area, the topography reflects closely the fold structure and topographically and structurally high areas coincide. Thick beds of sandstone and limestone form prominent ledges and cliffs or extensive dip slopes. Shaly rock units form steep slopes between the ledges of more resistant rocks.

SEDIMENTARY ROCKS

The sedimentary rocks exposed in the Comb Ridge-Navajo Mountain area range from Pennsylvanian to late Cretaceous in age. The thickness of the exposed stratigraphic section is between 6,500 and 7,000 feet. Pennsylvanian rocks crop out only in the canyons of the San Juan River east and west of Mexican Hat, Utah. Progressively younger rocks crop out to the south and west. Broad exposures of Permian and Triassic rocks are limited to the Monument Valley area and rocks of Jurassic age underlie the broad, high structural plain surrounding Monument Valley. The geologic map (Pl. 1) shows the areal distribution of these rock units.

The subsurface stratigraphic section consists of from 1,200 to 1,500 feet of rocks ranging from Pennsylvanian to Cambrian in age. Silurian and Devonian rocks are not known to be present. The stratigraphic relationships of the several formations are involved and will not be treated here. For detailed discussions on the subsurface stratigraphy the reader is referred to Bass (1944), Cooper (1955), Turnbow (1955), and Wengerd and Strickland (1954). The terminology used in Figure 2 follows that of Cooper (1954).

The nomenclature of the surface stratigraphic section has undergone repeated revision and the reader is referred to the following papers for detailed discussions: Baker and Reeside (1929), Baker (1936), Baker, Dane, and Reeside (1936

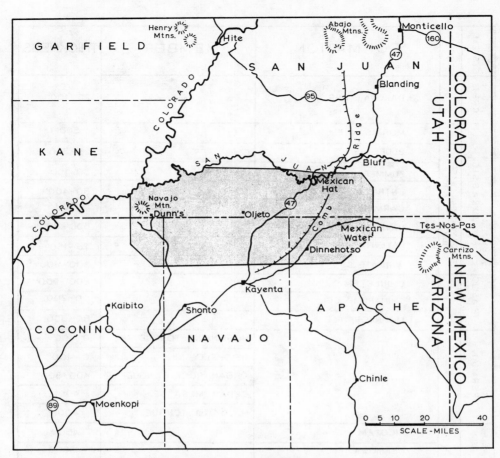

Fig. 1.—Index map showing location of Comb Ridge-Navajo Mountain area.

and 1947), Gregory (1917), Harshbarger, Repenning, and Irwin (1957), Repenning and Page (1956).

The nomenclature used in this report follows that recommended by Baker, Dane, and Reeside (1936) inasmuch as the stratigraphic units correspond closely to recognizable rock structure units. The stratigraphic section is shown in Figure 2. The thicknesses of several of the rock units vary considerably over the mapped area and the values shown are approximate.

Paradox formation (Pennsylvanian).—The Paradox formation is the oldest formation cropping out in the mapped area. Only the upper member is exposed and outcrops are restricted to the lower walls and floors of the canyons where the San Juan is incised across the Raplee and Halgaito anticlines (Fig. 4) near Mexican Hat, Utah. The exposed portion of this member consists of 200–500 feet of thin-bedded, black shales interbedded with gray-brown limestones and dolomites. The member marks the transition between the overlying Hermosa formation and the Salt member of the Paradox.

Hermosa formation (Pennsylvanian).—The Hermosa formation conformably overlies the Paradox formation and consists of about 1,200 feet of gray, fossiliferous limestones interbedded with a few thin beds of gray fine-grained sandstones and black shales. Exposures of this formation are limited to canyon walls of the San Juan River east and west of Mexican Hat, Utah. Systematic jointing is well displayed adjacent to the river where the Hermosa section is accessible.

Rico formation (Permo-Pennsylvanian).—The Rico formation consists of about 450 feet of thin-bedded to massive, light gray and red sandstones interbedded with light gray fossiliferous lime-

SYSTEM	GROUP	FORMATION	MEMBER	THICKNESS
CRET.		DAKOTA (?) SS.		25'-100'
JURASSIC		MORRISON FM.		500'+
JURASSIC	SAN RAFAEL GROUP	BLUFF SS.		25'-350'
JURASSIC	SAN RAFAEL GROUP	SUMMERVILLE FM.		80'-150'
JURASSIC	SAN RAFAEL GROUP	ENTRADA SS.		80'-400'
JURASSIC	SAN RAFAEL GROUP	CARMEL FM.		130'±
TRIAS. (?)	GLEN CANYON GROUP	NAVAJO SS.		500'-800'
TRIAS. (?)	GLEN CANYON GROUP	KAYENTA FM.		75'-150'
TRIAS. (?)	GLEN CANYON GROUP	WINGATE SS.		300'-400'
TRIASSIC		CHINLE FM.		800'-900'
TRIASSIC		SHINARUMP CGL.		50'-200'
TRIASSIC		MOENKOPI FM.		80'-300'
PERMIAN		CUTLER FM.	HOSKINNINI TONGUE	0'-100'
PERMIAN		CUTLER FM.	DE CHELLY SS.	0'-400'
PERMIAN		CUTLER FM.	ORGAN ROCK TONGUE	400'-800'
PERMIAN		CUTLER FM.	CEDAR MESA SS.	400'±
PERMIAN		CUTLER FM.	HALGAITO TONGUE	400'±
PERMO-PENN.		RICO FM.		450'±
PENN.		HERMOSA FM.		1200±
PENN.		PARADOX FM.		1500'±
PENN.		PINKERTON TRAIL FM.		50'-200'
PENN.		MOLAS FM.		25'-200'
MISS.		LEADVILLE LS.		400'-500'
MISS.		MADISON FM.		400'-500'
DEVONIAN		OURAY LS.		200'-400'
DEVONIAN		ELBERT FM.		200'-400'
DEVONIAN		ANETH FM.		200'-400'
CAMBRIAN		LYNCH DOL.		300'-1000'
CAMBRIAN		BOWMAN-HARTMAN LSS.		300'-1000'
CAMBRIAN		OPHIR FM.		300'-1000'
CAMBRIAN		TINTIC SS.		300'-1000'
PRECAMBRIAN				

FIG. 2.—Chart showing stratigraphic nomenclature used in this paper. Thickness shown in feet.

stones and red and purple shales. This formation conformably overlies the Hermosa and the contact is transitional. The Rico formation forms the upper walls of the canyons of the San Juan River east and west of Mexican Hat, Utah. Systematic jointing is well displayed in this formation where it is accessible.

Cutler formation (Permian).—The Cutler formation is subdivided into five members which are (in ascending order), the Halgaito tongue, the Cedar Mesa sandstone member, the Organ Rock tongue, the De Chelly sandstone member and the Hoskinnini tongue. This formation crops out primarily in Monument Valley and has limited exposures in the deep canyons west of Navajo Mountain.

Halgaito tongue.—The Halgaito tongue is the basal member of the Cutler formation and consists of about 400 feet of interbedded dark red silty sandstones and shales with a few thin beds of gray limestone. This member forms steep slopes below the cliffs of Cedar Mesa north of the San Juan River and along the crest of Halgaito anticline. Jointing in this member is exceptionally clear.

Cedar Mesa sandstone.—The Cedar Mesa member of the Cutler formation consists of about 400 feet of light gray to buff massive, cross-bedded sandstones with a few thin lenses of light gray limestone. The Cedar Mesa sandstone is best displayed along the cliffs north of the San Juan River and west of Mexican Hat, Utah. Toward the south and east the sandstones become less massive and grade into a redbed sequence similar to that of the overlying Organ Rock member. Sandstones of the Cedar Mesa form extensive dip slopes in the area extending from Comb Ridge on the east to Copper Canyon syncline and Organ Rock monocline on the west. Systematic jointing is well displayed where this member is exposed.

Organ Rock tongue.—The Organ Rock tongue of the Cutler formation consists of 400–800 feet of interbedded red-brown sandstones, mudstones, red and purple shales, and blue-gray limestones. The gross variations in thickness of this member may be due in large part to interfingering with the Cedar Mesa and De Chelly members. The Organ Rock tongue commonly forms steep slopes where it is exposed below the De Chelly cliffs and broad areal exposures are limited. The minimum thickness is found at the mouth of Nokai Creek northwest of Oljeto and the member thickens at the south and east within the mapped area.

De Chelly sandstone.—The De Chelly sandstone member of the Cutler formation consists of a trace to about 400 feet of massive, light buff to pink, medium- to fine-grained, cross-bedded sandstone. This massive sandstone forms prominent cliffs in the Monument Valley area and crops out extensively along the flanks of the Capitan anticline and Organ Rock monocline. Jointing is particularly well displayed along cliff faces in this rock unit (Fig. 19) but broad areal exposures are limited.

Hoskinnini tongue.—The Hoskinnini tongue of the Cutler formation consists of from 0 to 100 feet of even-bedded, red-brown sandstones and mudstones similar to those of the underlying De Chelly and Organ Rock members. Recent work by Stewart (1959) indicates this member may be basal Moenkopi. In outcrop pattern and topographic expression, the Hoskinnini tongue is essentially continuous with the underlying De Chelly sandstone.

Moenkopi formation (Early Triassic).—The Moenkopi formation overlies the Cutler formation conformably and, within the mapped area, consists of about 80–300 feet of thin, even-bedded, red-brown sandstones interbedded with dark red-brown shales. This formation forms slopes between ledges of overlying Shinarump conglomerate and the cliffs of the underlying Permian sandstones. The outcrop pattern and areal distribution is similar to that of the De Chelly sandstone.

Shinarump conglomerate (Late? Triassic).—The Shinarump conglomerate overlies the Moenkopi formation unconformably and consists of 50–200 feet of gray, coarse-grained, lenticular sandstone with interbedded lenses of pebble conglomerate and light green shales. This formation caps mesas and forms extensive dip slopes south and west of Monument Valley. The broad exposures of the Shinarump make it extremely useful in reconstructing the joint pattern in this area.

Chinle formation (Late Triassic).—The Chinle formation overlies the Shinarump conglomerate conformably and consists of 800–900 feet of brilliant variegated shales interbedded with massive, reddish brown, lenticular, cross-bedded sandstones in the upper part of the section. The upper, sandy part of the Chinle is now included as a member (Red Point) of the overlying Wingate sandstone (Harshbarger, Repenning, and Irwin,

1957). The Chinle forms a steep slope everywhere between the dip slopes and ledges of the underlying Shinarump and the massive cliffs of the overlying Wingate. The areal extent of exposures is limited.

Glen Canyon group.—The Glen Canyon group consists of an upper unit (Navajo) and a lower unit (Wingate) of massive, cross-bedded sandstones separated by a relatively thin unit (Kayenta) of lenticular sandstones, mudstones, and shales.

Wingate sandstone (Triassic?).—The Wingate sandstone overlies the Chinle formation conformably and the contact between the two is gradational. The Wingate consists of 300–400 feet of light reddish brown, fine-grained, cross-bedded sandstone. This unit corresponds with the Lukachukai member of the Wingate as defined by Harshbarger, Repenning, and Irwin (1957). The Wingate forms prominent vertical scarps above the Chinle slope throughout the mapped area. Jointing is well displayed in the cliff faces but the areal extent of exposures is limited.

Kayenta formation (Jurassic).—The Kayenta formation overlies the Wingate sandstone conformably and the contact between the two is gradational. The Kayenta formation consists of 75–150 feet of buff to reddish brown, fine-grained, lenticular, cross-bedded sandstones interbedded with red to gray mudstones and shales. This formation commonly forms a gentle slope between the cliff-forming Wingate and Navajo sandstones.

Navajo sandstone (Jurassic).—The Navajo sandstone overlies and intertongues with the Kayenta formation. The Navajo consists of 500–800 feet of massive, pale reddish brown, fine-grained, cross-bedded sandstone. Thin, lenticular beds of gray, non-fossiliferous limestone occur in the upper half of the formation.

The outcrop area of the Navajo sandstone is greater than that of any other formation within the mapped area. The lithologic uniformity, extensive exposures and wide areal distribution provide ideal conditions for the study of systematic joints and more than half the joint data used in the reconstruction of the regional joint pattern come from this formation.

San Rafael group (Middle to Late Jurassic).—The San Rafael group within the mapped area consists of, in ascending order, the Carmel formation, the Entrada sandstone, Todilto limestone,

Summerville formation and Bluff sandstone. Exposures of formations in this group are limited to the eastern and western margins of the mapped area and the several formations were not differentiated in the field. The Carmel formation is separated from the Navajo sandstone by an erosional unconformity.

Only the Carmel and Entrada formations are present at Navajo Mountain where they are largely obscured by the extensive talus slopes on the mountain flanks.

All formations of the group are recognized in the eastern part of the mapped area at Garnet Ridge northwest of Dinnehotso, Arizona, and in the slopes of the mesas north of Mexican Water, Arizona.

Morrison formation (Late Jurassic).—The Morrison formation forms the caprock of mesas east of Comb Ridge. A complete section is present only on Navajo Mountain, where it is obscured by extensive talus slopes. The formation consists of 500–1,200 feet of interbedded, gray lenticular sandstones grading upward into varicolored shales and buff to reddish, lenticular sandstones.

Dakota(?) sandstone (Late Cretaceous).—A gray-white, silicified, conglomeratic sandstone caps Navajo Mountain. This has been identified tentatively as Dakota (?) (Baker, 1936) but may be the uppermost sandstone of the Morrison formation.

Unconsolidated deposits (Late Cenozoic).—Unconsolidated alluvial deposits consisting of fine-grained, cross-bedded lenses of sand interbedded with lenses of gravel are found as fill in the dry washes and canyons of the area.

Active and inactive longitudinal dunes are a common feature throughout the area and a thin cover of loose sand is present almost everywhere.

IGNEOUS ROCKS

Igneous rocks in the Comb Ridge-Navajo Mountain area are represented by minette dikes and "plugs" that are present along the length of Comb Ridge and in the Oljeto and Mexican Hat areas. These intrusives belong to the Navajo volcanic field and have been dated as "Tertiary" (Gregory, 1917; Baker, 1936). Stratigraphic evidence in the Hopi Buttes volcanic field at the south (Shoemaker, 1953) indicates volcanic activity in that area as late as Pliocene time. The intrusives in the mapped area may be as young but stratigraphic evidence is lacking as the youngest formation disturbed by igneous activity is of Morrison age.

"Plugs."—The large plug-like masses such as Agathla and Boundary Butte are composed largely of a network of anastomosing dikes with the spaces between the dikes filled with rock fragments of a wide range of size and composition derived from lower in the stratigraphic section.

Dikes.—Dikes appear both as separate features and in close association with the plugs. Without exception, the dikes have been intruded along joints without disrupting the adjacent wall rock. Maximum alteration of the rock occurs along joints directly above a dike and some dikes may be traced thus for some distance after they have disappeared under rock cover. Widened portions of dikes contain rock fragments derived from lower in the section and similar to those found in the plugs.

REGIONAL STRUCTURE

The extent and disposition of the major structural elements of the Colorado Plateau tectonic province is shown in Figure 3. The mapped area is in the south-central part of the province.

The upwarps and basins of the Colorado Plateau are markedly asymmetrical. Commonly, a long monoclinal flexure defines sharply the transition between a basin and the juxtaposed upwarp. The opposing flank of the upwarp or basin consists of a long, gentle structural slope interrupted by more or less well defined subsidiary folds. The axes of the upwarps and basins lie near to and follow the trends of the monoclines. The axes of the subsidiary folds are ordinarily offset *en échelon* and follow closely the trend of the larger folds.

In general, structural relief is greatest in the north and northeast part of the Plateau and decreases southward, reaching minimum in the San Juan and Black Mesa basins. Maximum deformation within the borders of the tectonic province is localized along the short, steep flanks of the larger folds. Exceptions to this generality are found in the salt anticlines of the Paradox Fold Belt. Faults of great displacement are largely restricted to the borders of the Plateau.

Strong northwest fold trends predominate over the Plateau even though several of the larger structural elements such as the San Rafael and Monument upwarps and the San Juan and Uinta basins have north-south, northeast, or east-west

trends. The disposition of the major fold trends with respect to trend direction does not appear systematic. Fold asymmetry, however, does show a peculiar regional distribution. With few exceptions, folds northeast of a line drawn from the north flank of the Zuni upwarp to the north end of the San Rafael upwarp are asymmetric toward the south or southwest and folds southeast of this line are asymmetric toward the east or northeast. Kelley (1955, Fig. 11) recognizes also a basic difference in structural geometry across this line and has divided the Plateau into "Eastern" and "Western" tectonic subprovinces.

STRUCTURE OF COMB RIDGE-NAVAJO MOUNTAIN AREA

The Comb Ridge-Navajo Mountain area comprises parts of three major structural divisions of the Colorado Plateau tectonic province; the Monument upwarp, Henry Mountains basin, and the Circle Cliffs upwarp (Figs. 3, 4).

The most prominent fold in the map area is the Monument upwarp which trends generally north-south. The southern third of the upwarp, extending from Comb Ridge monocline on the east to the Oljeto syncline on the west was investigated in this study. Within the area the upwarp swings from north-south to southwest and terminates in the vicinity of Kayenta, Arizona. It is a broad, asymmetrical anticlinal arch with a short, steep east limb and a long, gentle west limb interrupted by several prominent subsidiary folds (Fig. 4).

East of Comb monocline the strike of the folds changes abruptly to northwest. These folds represent the north end of the Defiance upwarp.

North of the state line between Arizona and Utah the gentle flank of the Monument upwarp merges with the south end of the Henry Mountains basin represented by the Nokai syncline. West of the Nokai syncline the Balanced Rock monocline forms the east flank of the southern end of the Circle Cliffs upwarp. The trend of these folds is north to northwest.

South of the state line, the prominent Organ Rock monocline marks the transition between the Monument upwarp and the structurally high area on the west.

The structural dome of Navajo Mountain is on the western border of the mapped area and is peculiar because of its relatively isolated position and because it does not follow the regional structural grain. This oval fold strikes northeast and is

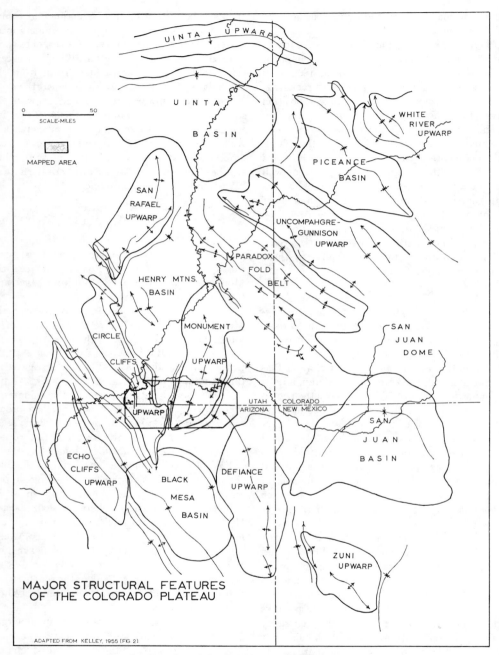

FIG. 3.—Location of Comb Ridge-Navajo Mountain area with
respect to major structural features of Colorado Plateau.

asymmetric toward the northwest (Baker, 1936, Pl. I). It is inferred (Gregory, 1917; Baker, 1936; Hunt, 1953) that the dome is the result of the intrusion at depth of a stock similar to those found in the Henry Mountains on the north. No igneous rock crops out, however, and it is problematical whether the dome is truly analogous with the Henry Mountains.

In the mapped area the folds are asymmetric and, with the notable exceptions of the Raplee

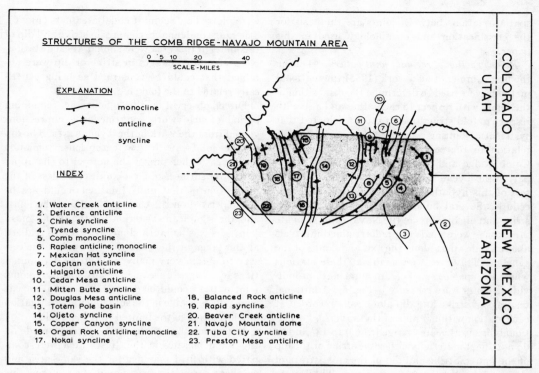

STRUCTURES OF THE COMB RIDGE-NAVAJO MOUNTAIN AREA

0 5 10 20 40
SCALE-MILES

EXPLANATION

monocline

anticline

syncline

INDEX

1. Water Creek anticline
2. Defiance anticline
3. Chinle syncline
4. Tyende syncline
5. Comb monocline
6. Raplee anticline; monocline
7. Mexican Hat syncline
8. Capitan anticline
9. Halgaito anticline
10. Cedar Mesa anticline
11. Mitten Butte syncline
12. Douglas Mesa anticline
13. Totem Pole basin
14. Oljeto syncline
15. Copper Canyon syncline
16. Organ Rock anticline; monocline
17. Nokai syncline

18. Balanced Rock anticline
19. Rapid syncline
20. Beaver Creek anticline
21. Navajo Mountain dome
22. Tuba City syncline
23. Preston Mesa anticline

COLORADO
UTAH

NEW MEXICO
ARIZONA

FIG. 4.—Location of folds in Comb Ridge-Navajo Mountain area.

monocline and Navajo Mountain, the steep limbs of the folds dip east or northeast. Dips rarely exceed 5° except locally along the steep limbs of folds (monoclines) where dips range as high as 30°. Faulting is limited to high-angle normal faults of a few feet displacement, that, in most examples, follow large joints.

Two groups of diatremes disturb the strata along Comb Ridge. One group occurs on the west side of the scarp south of the San Juan River and the other at Garnet Ridge near the crest of Comb Ridge, a few miles northwest of Dinnehotso, Arizona (Pl. I).

PROCEDURES

Base map.—Adequate topographic maps covering the entire mapped area were not available in the summers of 1955 and 1956. A base map was constructed directly from standard 15′ quadrangle aerial mosaics, scale 1 inch equals 1 mile (Soil Conservation Project, Nav 35A). The coordinates used on the base map are those given on these mosaics.

Recent township surveys (1954) based on the Gila and Salt River meridian, Arizona, extend only into the eastern part of the mapped area. The township grid in this area was reconstructed on the base map from section corners recovered in the field and from township plats.

Geologic map (Pl. I).—Formational contacts were checked in the field and traced directly on the aerial mosaics. These contacts were then transferred to the base map. Extensive areas of wind-blown sand are shown also inasmuch as such areas prevent the observation of joints both in the field and on aerial mosaics. The position of fold axes is plotted from observations in the field and in the western part of the mapped area, with additional reference to the structure contour map by Baker (1936, Pl. I). The original map has been redrawn and simplified for publication.

The regional joint pattern is represented by a network of evenly spaced lines showing the direction and extent of the several elements of this pattern.

Individual joints.—The features of individual joints of all types were studied in detail throughout the area. Particular attention was paid to the nature of the joint traces and the surface features of joint faces. This study and the study of the spatial relations between joints are the basis for the classification and terminology used in this paper.

Reconstruction of regional joint pattern.—During the first summer of field work (1955) conventional methods were used to determine the disposition of the joints with respect to each other and also with respect to fold structures. It was anticipated that systematic jointing would show some definable relation to the axes of folds. The area east of Comb Ridge and along the flanks of the north part of the Tyende syncline was used for an initial test of this hypothesis and also to test field procedures. Several hundred determinations of the strike and dip of joints in the Navajo and Wingate sandstones were made at predetermined intervals about a mile apart along the axis and flanks of the syncline. Traverses were made at these stations where the spacing, strike, dip, and length of individual joints were measured (Fig. 5). Additional readings of strike and dip alone were recorded at each station within $\frac{1}{4}$ mile of the traverse. Formational dips in this area range from 0 to 12° and so do not affect materially either the strike or the dip of the joints. Strike and dip of the cross-strata of the sandstones were also obtained. These data

were plotted on Schmidt equal-area nets (Fig. 6). The range of definable changes in strike and dip of the joints in this area were similar at each station. No systematic changes in strike or dip were obtained that could be construed as being genetically related to the fold (Fig. 7).

Direct observation throughout the remaining period of field work confirmed the impressions gained from the work in the Tyende area. The dip of the joints, with few exceptions, remained within 25° either side of the normal to the upper and lower surfaces of a rock unit regardless of the disposition of that unit. Joint sets could not be differentiated on the basis of the dip of the joints and the significant criterion for the definition of joint sets was azimuth alone. It was also realized at this stage of the work that it would be necessary to obtain very many joint strike determinations over a great area before any sort of regional joint pattern could be defined. The aerial mosaic were indispensable in the subsequent investigation of a large area in the limited time available.

Determinations of the strike of joints made on the aerial mosaics in the "test" area were compared with field readings for the same area and excellent correlation was found. The various sets

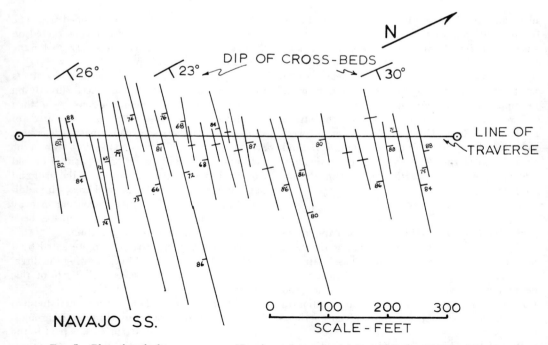

FIG. 5.—Plan of typical traverse across Navajo sandstone in vicinity of Mexican Water, Arizona.

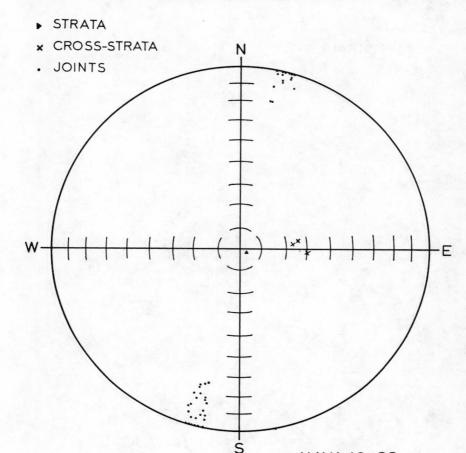

FIG. 6.—Data from traverse across Navajo sandstone plotted
on Schmidt equal area net (upper hemisphere).

defined in the field could be traced continuously on the mosaics from one field station to the next except in areas of thick sand cover.

Aerial mosaics were used extensively in the summer of 1956, particularly in the investigation of areas difficult of access. The only area where the mosaics were not a valuable aid was south of the San Juan River between the Comb monocline and the Organ Rock anticline. Here the joints of the thin-bedded Permian shales and sandstones do not show directly on the aerial mosaics and, in places, the sand cover is thick and extensive. All data in this area are based on field readings.

In a preliminary report (Hodgson, 1956), the joint data then available were summarized in the form of rose diagrams for each 7½′ quadrangle. The joint pattern was also reconstructed graphi-

cally by means of a network of lines showing the direction and extent of the joint sets. The second method presented a clearer picture of the regional pattern and the relation of the pattern to fold axes. It is the method of representation used in the present paper. The joint pattern shown on Plate I is the result of the compilation of all information pertaining to the areal distribution of joint sets of regional extent as derived from field observations and the study of aerial mosaics.

DEFINITION OF TERMS

Definitions of the term "joint," as well as several other terms pertaining to jointing are, in general, either descriptive or genetic.

The use of genetic terms implies that the origin of the joints has been determined. The use of such

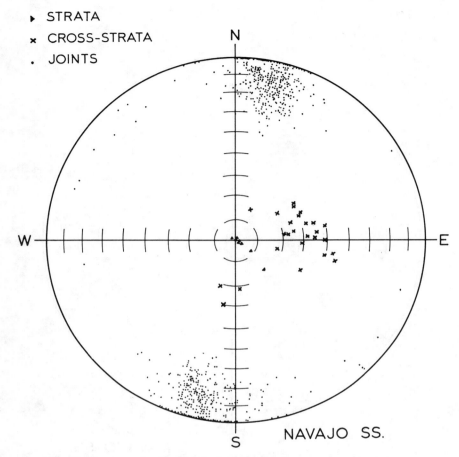

Fɪɢ. 7.—Summary of strike and dip data from twelve traverses on both flanks of
Tyende syncline between Mexican Water and Dinnehotso, Arizona.

terms is not justified unless the origin is known and may be misleading when applied to the observation and interpretation of classes of joints other than those to which they were originally applied.

The descriptive terms used in this paper are here defined.

Joint.—The term joint refers primarily to the actual fracture traversing a rock and represented by a fine line or "trace" that marks the intersection of the joint and rock surfaces. For practical use it must cover also the more abstract idea of a joint as represented by a cleft or fissure, the face of one side of the fracture as on a weathered outcrop, or some other diagnostic physical feature.

A joint may be defined formally as a fracture that traverses a rock and is not accompanied by any discernible displacement of one face of the

fracture relative to the other. As defined, the term joint is synonomous with the term "diaclase" (Daubree, 1879, p. 351).

The term joint alone is not specific and must be modified by some descriptive term so that the several varieties of jointing can be differentiated.

The joints in the mapped area can be classified conveniently into two categories on the basis of their pattern in plan and by distinctive features of individual joints. These categories are (1) systematic joints and, (2) non-systematic joints.

Systematic joints.—Systematic joints occur in sets. The joints composing each set are parallel or sub-parallel in plan and may or may not show a similar relation in section.

Joints of each set intersect (continue across

joints of other sets where more than one set is present).

Systematic joints show the following distinct structural features.

1. Systematic joints commonly exhibit planar or broadly curved traces or surfaces.

2. Systematic joints most commonly occur about at right angles to the upper and lower surfaces of the rock units in which they are present, regardless of the attitude of these units.

3. Surface structures on the faces of systematic joints are oriented.

4. Systematic joints cut across other joints.

Non-systematic joints.—Non-systematic joints display random rather than oriented patterns in plan and section. Such random, curvilinear patterns are the most distinguishing feature of this type of joint but there are additional features that are also diagnostic.

1. Non-systematic joints meet but do not cross other joints, whether systematic or non-systematic.

2. Non-systematic joints are generally strongly curved in plan (also in section in some rocks).

3. Non-systematic joints commonly terminate at bedding surfaces.

4. Surface structures on the faces of non-systematic joints are not oriented.

Cross-joints.—Cross-joints are a distinct variety of non-systematic joints. These joints extend across the intervals between systematic joints and are roughly normal to the systematic joints and to the upper and lower surfaces of the rock units. In detail, however, cross-joints have more irregular surfaces than systematic joints, show no oriented surface structures and commonly terminate at bedding surfaces and against other joints.

Joint trace.—A joint trace is the line formed by the intersection of a joint surface with an exposed rock surface.

Joint zone.—In plan, systematic joints commonly occur in narrow zones where the individual joints are only slightly offset from each other. These lines or "joint zones" are separated by the predominant unit spacing peculiar to the joint set.

Fissure and fissuring.—The superficial expression of a joint is commonly a fissure resulting either from mechanical separation of the faces of the joint or weathering along the joint surfaces. The term "fissuring" as used in this paper refers specifically to the process of development and ex-

tension of a fissure along joint surfaces by weathering agents.

Plan and section.—The map view of the upper surface of a rock unit regardless of the attitude of that unit, is termed the "plan view" or "plan." "Section" is the view at right angles to the plan view.

Terminal offset faces.—Terminal offset faces are the faces of small *en échelon* fractures that mark the edges of systematic and some non-systematic joints.

Structure rock unit.—The concept of a structure rock unit is fundamental to the discussion of the mechanics of jointing as well as of other structural phenomena.

The stratigraphic section is commonly divided and classified on the basis of mappable lithologic changes (A.G.I. Glossary, 1958, p. 114 for accepted definitions). The stratigraphic section may be divided also into units with synchronous boundaries based on geologic time (A.G.I. Glossary, 1958, p. 299).

None of these concepts of division of the stratigraphic section is completely satisfactory for the mapping or interpretation of structure, whether jointing, folding, or faulting.

Even though rocks may appear radically different at a weathered outcrop, they may be very similar in a structural sense, that is, in reaction to stress. A structure rock unit might be equivalent to a geologic formation or be smaller or larger, according to the type of rock and the stresses involved.

A structure rock unit is thus a rock unit defined with respect to its reaction to external and (or) internal stresses. It may range in thickness from a single stratum to a series of strata several hundreds or thousands of feet thick. Thus structure rock units related to jointing conceivably differ in thickness and rock type from those pertaining to folding or faulting in the degree that the stresses involved in producing these structures may differ.

At present it is difficult to use this concept in a quantitative sense because of the inability to define precisely discrete structural units, particularly with respect to jointing. In a general sense, the stratigraphic rock units defined by Baker (1936) appear different enough lithologically to be classed as gross structure rock units. For example, the jointing in the Navajo sandstone is of a different order of magnitude than that in the relatively thin Shinarump conglomerate. The sandstones are of different thickness and geometry and these features have some effect on the spacing and size of the joints but, beyond this, little can be said at present. The concept of "structure rock units" is believed useful, however, and is used qualitatively in the following discussion.

SPATIAL RELATIONS OF SYSTEMATIC JOINTS

The spatial relations of systematic joints are of primary importance as they are the context within which theories concerning the origin of joints must be applied. In fact, these spatial relations place somewhat severe limitations on theory.

JOINT TRACES

A joint trace is a thin line marking the intersection of a joint surface and an exposed rock surface. Regardless of length, the traces of systematic joints show similar features that are found both in plan and section.

When effects of topography are removed, the trace is found to be a straight or gently curved sigmoidal line. The observed lengths of individual traces range from a few inches to more than 400 feet.

A significant feature of the joint trace is the manner in which it terminates. A trace tends to become irregular near either end and the ends of many traces curve abruptly from the line of strike, commonly, but not invariably, toward the nearest adjacent joint of the same set. Some curved ends terminate against an adjacent systematic joint which, in turn, may terminate against the first joint in a similar manner. The direction of curvature may be left- or right-handed, or both, at the ends of a single trace. It is characteristic that these deviations are not systematic.

Many joint traces bifurcate near their terminations. Where a single trace splits, the two resultant traces tend to curve away from each other and extend a few inches or feet beyond the single trace. The angle between the bifurcating traces is not constant and ranges from about 20° to 160°.

Where the ends of joint traces are examined on fresh rock surfaces, it is apparent that they become increasingly weak in expression toward the terminus until they are no longer discernible. It may not be possible to determine precisely where a trace terminates unless the joint has undergone some weathering. Figure 8 shows the several com-

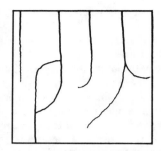

FIG. 8.—Diagrammatic plan view showing terminations of systematic joints.

mon types of terminations of systematic joint traces.

EXTENT OF JOINTS

The extent of a joint can be inferred from its trace in plan and section or determined by direct observation of the joint face where it has been freshly exposed. Such observations show that joints range in size from a few square inches to several hundred square feet.

Joints appear to be roughly equidimensional or, particularly in thin-bedded rocks, elongate in the direction of the axis of plumose surface structures (see section on plumose structures).

Joints in thin-bedded rocks commonly extend across many layers of strata and appear, in many examples, to be related to the gross lithologic character of the section rather than to an individual stratum.

Relatively few joints, however, extend completely through very thick rock units such as the Navajo, Wingate, or De Chelly sandstones. Many joints must be contained entirely within such units and are recognizable only where one end of the joint has been exposed by erosion in a canyon or cliff.

Some joints extend across the boundaries between very distinct lithologic types. This is shown clearly in Figure 9 where the larger joints cut across the boundary between the De Chelly sandstone and the thin-bedded shales and sandstones of the Organ Rock tongue.

PARALLELISM IN PLAN

The most prominent characteristic of systematic joints is their parallelism when viewed in plan. The joints do not occur as single fractures but are grouped into sets within which the strike of each joint is parallel or sub-parallel with others

FIG. 9.—View west across Organ Rock monocline near Oljeto, Arizona. Differences in spacing between systematic joints in De Chelly sandstone and thin-bedded Organ Rock strata are shown clearly. Some larger joints in De Chelly cut across boundary between two members.

of the same set. Each such set in turn appears to be a distinct unit with identifiable boundaries in plan. Such a unit commonly is of considerable areal extent and is distinguished from other sets by the unique strike that it maintains.

Whether or not the joints of a set appear truly parallel in plan or tend to deviate to a greater or less extent from the mean strike of the set appears to be related primarily to lithologic character. Exceptions occur, but it may be stated as a general rule that the joints are most nearly parallel with each other in homogeneous rocks that are evenly bedded such as massive siltstones, some limestones, and massive, homogeneous, cross-bedded sandstones. Joints in coarse-grained, lenticular or poorly bedded rocks commonly show a greater departure from the mean strike of a set than those in the fine-grained homogeneous rocks.

PARALLELISM OF JOINTS IN SECTION

The parallelism so well displayed in plan by a set of systematic joints ordinarily appears much less perfect when the joints are viewed in section. In the mapped area, dips of systematic joints commonly range from 0 to 25° either side of the normal to the upper and lower surfaces of a rock unit, and in thick massive beds may depart as much as

45°–50°. The range in dip for joints in the Navajo sandstone in the vicinity of Mexican Water, Arizona, is shown in Figure 7.

There is no apparent system of the direction in which the joints of a set may dip. Very locally the dip may be predominantly in one direction, as observed in several beds of sandstone in the Halgaito near the San Juan River at the Raplee monocline. Over any distance, however, the joints appear to dip equally in either direction.

The magnitude of dip of joints of a particular set may vary from one rock unit to another through the section. The range in dip commonly is greater in thick, massive sandstones than in flat-lying, thin-bedded rocks. Comparison of the dips of the exceptionally perfect joints in the mudstones and shales of the Halgaito tongue (Fig. 10) with those of the Wingate and Navajo sandstones (Fig. 11) shows that in the Halgaito section the joints are almost equally parallel in plan and section.

As the range of dips for all systematic joints in a rock unit is similar, the dip of the joints can not be used as a criterion for differentiating one set from another.

ZONAL DISPOSITION OF JOINTS

The traces and exposed surfaces show that, within each joint set, joints are arranged in narrow zones within which individual joints are only slightly offset from each other. These zones are separated by the predominant unit spacing peculiar to that set and rock unit.

A characteristic feature of this zonal arrangement is that the direction of offset from joint to

FIG. 10.—Systematic joints and cross-joints in massive siltstones of Halgaito about 1 mile east of Mexican Hat, Utah. San Juan River in foreground. Height of outcrop about 20 feet.

FIG. 11.—View east of Segihatsosie Canyon about 10 miles north of Kayenta, Arizona. Wingate sandstone in foreground and Navajo sandstone forming massive cliff in background. Differences of spacing of systematic joints are shown in the two formations and relatively low dip of large joint in Navajo cliff, right center of photograph.

joint along a zone is not systematic in either plan or section. Within a set or zone, the offsets occur with approximately equal frequency in either direction, some in opposite directions at either end of an individual joint. Figures 12a and 12b show schematically the zonal arrangement and the mode of offset of systematic joints.

SPACING OF JOINTS

Examination of joints in plan and section over the mapped area showed the following recognizable variations in the spacing of joints.

1. Local departures from the average spacing of joint zones in a single set.
2. Variations in average spacing of joint zones from set to set in the same area and rock unit.
3. Variations in average spacing of joints of the same set in rock units of differing thicknesses and lithologic character in the same area.
4. Irregular areas where systematic jointing is non-existent or poorly developed.

It is quite possible that there is more than one order of systematic joints in some of the thicker rock units, particularly those composed of thin, flat-bedded units; one order of joints being related to the gross structure rock unit and other orders

to smaller units. No systematic study was made of this particular problem.

Recognition of the complex problems of spacing and the difficulty of defining a structure rock unit, precluded any attempt in this study to establish a precise mathematical relation between the thickness of rock units and the spacing of joints. In general, however, the spacing of joints is related to the thickness of rock units in that all systematic joints tend to have a wider spacing in thick, massive rock units than in thin ones. Figure 13 shows schematically, observed variations in the spacing of systematic joints.

INTERSECTION OF JOINTS

Joints of any set will intersect joints of other sets. In plan, the angle of intersection between any two sets ranges from less than 15° to 90°, with the angle of intersection between two sets remaining essentially constant over the area. Deviations from the average angle of intersection are comparable with deviations of individual joints from the average strike of a joint set.

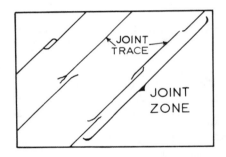

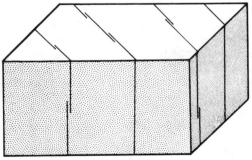

FIG. 12a (Upper). Diagrammatic plan view showing arrangement of systematic joints in narrow zones of *en échelon* fractures (joint zones).
12b (Lower). Schematic block diagram showing joint zones.

Joints of the *same set* may intersect in section. This relationship is observed in the Navajo sandstone with unusual clarity and is shown schematically in Figure 14a. Such intersections in section, however, appear to be rare except in the thick, massive sandstones of the region, for there is less tendency for systematic joints to deviate greatly from the normal to the bedding surfaces in thin, flat-bedded rocks.

The joints of different sets also intersect in section. Again, this relation is observed most commonly in thick-bedded rocks (Fig. 14b).

Each joint set could be considered as a separate entity in that the physical characteristics of any set are not discernibly modified by the presence of other joint sets.

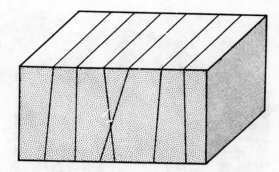
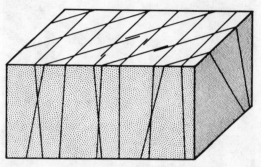

FIG. 14a (Upper).—Schematic block diagram showing manner in which joints of same set intersect in section. 14b (Lower).—Schematic block diagram showing manner in which joints of different sets intersect in section. Regular pattern observed in plan is not found in section.

RELATION OF SYSTEMATIC JOINTS TO CROSS-BEDDING

Systematic joints commonly are nearly perpendicular to the upper and lower surfaces of sedimentary rock units. It is important to consider whether they remain perpendicular where the bedding of the rock unit is not parallel with these surfaces, as in the prominent cross-beds of the Navajo, Wingate, and De Chelly sandstones. The dip and strike of both joints and cross-beds were determined in the Comb Ridge area and plotted on Schmidt nets for comparison (Fig. 7). Direct observations on the effect of cross-bedding on jointing were made throughout the course of the field work (Fig. 15). The presence of cross-bedding does not appear to have any influence on the form or orientation of the systematic joints. With few exceptions, also noted in other rock types, the joints occur nearly normal to the upper and lower surfaces of the gross sandstone unit. Such sandstones are structure rock units with respect to systematic jointing.

Departures from the average spatial relations appear to be related more clearly to obvious lithologic inhomogeneities in rock units such as lenticular bedding, and sharp lithologic changes rather than bedding of any type in rocks of homogeneous composition.

SPATIAL RELATIONS OF NON-SYSTEMATIC JOINTS

Non-systematic joints of several types extend across the intervals between systematic joints and characteristically *terminate against* systematic joints or other, non-systematic, joints. The angle of juncture between a systematic and non-systematic joint commonly approaches a right angle, whereas the angle of juncture between two non-systematic joints is highly variable.

Non-systematic joints, although curved in plan, are almost invariably normal to the bedding in thin-bedded rock units. They may cross several thin beds that comprise a distinct lithologic unit.

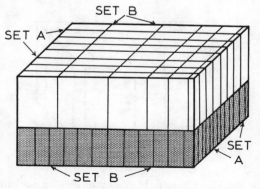

FIG. 13.—Schematic block diagram showing observed variables in joint spacing.

FIG. 15.—View east across Todecheenie Bench about 15 miles north of Kayenta, Arizona. Geometric relation between systematic joints and cross-strata of De Chelly sandstone is clearly shown. Same set of joints is present also in Shinarump conglomerate, capping mesa.

Figure 16 illustrates clearly the relation in plan between systematic and non-systematic joints.

Non-systematic joints of large dimensions are found in the thick, massive cross-bedded sandstones of the region and appear to be restricted to such units. They are, however, difficult to observe

clearly as they are rarely fissured by weathering. In form they appear similar to smaller-scale non-systematic joints except that they are equally curved and irregular in both plan and section.

CROSS-JOINTS

Cross-joints are joints that are superficially similar to systematic joints in having a relatively straight trace in plan but differ from them in several other important respects: (1) cross-joints do not intersect systematic joints, (2) cross-joints commonly terminate against prominent bedding surfaces in section, and (3) although some of their surfaces approach a plane as do those of systematic joints, they are more typically irregular, sinuous or rough.

Figure 10 clearly illustrates the relations between cross-joints and systematic joints in a massive siltstone of the Halgaito east of Mexican Hat, Utah, where only one set of joints is present. Joints belonging to this set are of large size, planar, and cut cleanly across the bedding surfaces of the rock. In contrast, the cross-joints, although nearly at right angles to the systematic joints and normal to the bedding surfaces or the rock, are clearly more irregular in form and surface detail. Most important, however, is the fact that the cross-joints invariably terminate *against* the systematic joints. Also, the cross-joints terminate against prominent bedding surfaces, some being restricted to individual beds and others crossing several thin beds. These relations are shown diagrammatically in Figure 17.

In section the differences between systematic joints and cross-joints are readily apparent in an ideal exposure. In plan, and in less perfect expo-

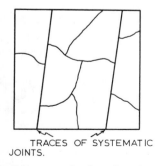

TRACES OF SYSTEMATIC JOINTS.

FIG. 16.—Diagrammatic plan view showing typical pattern of non-systematic joints and their characteristic termination against systematic joints.

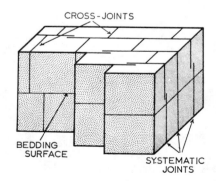

FIG. 17.—Schematic block diagram showing relations between cross-joints and systematic joints and between cross-joints and prominent bedding surfaces.

sures, they can be identified only by their traces, the most important criterion in such cases being whether or not joint traces cross each other or whether the joints of one "set" invariably terminate against joints of another set (Figs. 16 and 17).

The superficial similarity between cross-joints and systematic joints makes it imperative to differentiate them so that the cross-joints are not mistaken for members of a non-existent set.

FEATURES OF JOINT SURFACES

The discussion in the foregoing section has been limited to a consideration of the spatial relations of joints. The character of the individual joint, however, is equally important, and exhibits certain features that must be considered in any hypothesis of the origin of joints.

The writer's concept of the nature of a complete joint surface is based on the observation of many such surfaces in the field. A complete, fresh joint surface of any magnitude rarely, if ever, is entirely exposed. A composite must be pieced together from the observation and comparison of many such surfaces.

In addition, the detailed markings of joint surfaces are commonly difficult to detect under the most favorable conditions because of their low relief and fine texture. Even on a freshly exposed joint surface, the lighting must be in the right direction to show the features clearly and, they are seldom visible under any conditions on surfaces exposed to the direct attack of weathering for any length of time.

Systematic joints are, in general, planar fractures. Figure 18 shows a weathered joint surface in the Shinarump conglomerate. The surface features have been destroyed by erosion but the planar nature of the fracture is apparent. An undetermined part of the surface, however, has been removed or modified by erosion. The upper part of the joint surface merges into the rounded forms of the mesa top and the lower part of the surface extended at one time into the underlying shales which have been partly removed by weathering. Little is disclosed besides the over-all form of the joint.

Figure 19 illustrates slightly less weathered joint surfaces of great size in the De Chelly sandstone. The face of the cliff shows offsets or steps resulting from several closely spaced joints arranged *en échelon* (joint zone). No individual joint

FIG. 18.—View northeast from rim of Hoskinnini Mesa near Oljeto, Arizona, showing planar nature of systematic joint in Shinarump conglomerate.

is exposed completely, however, and each surface disappears under an adjacent slab at one end. Weathering has obscured details of the joint surfaces.

PLUMOSE STRUCTURE

The freshly exposed face of a systematic joint may show a characteristic structure composed of ridges of very low relief. The pattern displayed by these ridges is similar to that formed by placing two feathers or plumes in a straight line with the butt ends joined. These markings are here called plumose structures, the term "plume" being used first by Parker (1942, p. 396) in describing similar structures. The term "feather fracture" is used in reference to these structures by Woodworth (1897, p. 165) but this term should be dropped despite priority because of possible confusion with the term "feather joint" (Billings, 1954, p. 117).

Woodworth's description of the details of plumose structures is set forth clearly and is remarkably complete. The joint surfaces described by him are of small size not exceeding 200 mm. in length but his observations agree in detail with those made in the present study.

Parker (1942, p. 396 and Pl. 3) describes plumose structure noting the oriented nature of the plumes on systematic joints and the fact that some plumes extend over several rock layers, particularly in shales.

FIG. 19.—View southeast from rim of Hoskinnini Mesa about 5 miles southwest of Oljeto. Stepped surfaces of *en échelon* systematic joints are exposed in face of massive cliff of De Chelly sandstone. No individual joint surface is completely exposed.

Raggatt (1954, Fig. 1) shows a variety of plumose structure on the faces of joints in a siltstone member of the Demon's Bluff formation. From the illustration these appear to be similar to the "rays of percussion" of Woodworth (p. 166) and can be seen also on the face of a systematic joint in the right center of Figure 20.

With few exceptions, the faint ridges of the plumose structure curve outward from a central line or axis which ordinarily occurs either parallel with, or normal to, the upper and lower boundaries of a rock unit. Plumose structures with both orientations are found on the faces of joints belonging to the same set but are observed most commonly parallel with the boundaries of rock units. No plumes observed on systematic joints had orientations other than the two cited.

The relief of the ridges delineating the plumose pattern is proportional to the size of the joint surface and is greater on large surfaces than small.

Figure 20 shows plumes on joint surfaces in a massive sandstone of the Halgaito. The pattern is particularly clear on the joint surface in the center of the photograph. The axis of the plume is parallel with the bedding surfaces in the rock and lies near the top of the outcrop. Individual linear ridges curve sharply downward and away from the axis, diverging from each other and forming an increasingly coarse pattern toward the edge of the joint.

Observations on many such surfaces indicate that a plume is a roughly symmetrical structure and that the axis of the plume occurs near the center of a joint surface (not necessarily at the intersection with a bedding surface). By applying this generalization to the surface in Figure 20, it appears that about half the joint surface has been removed by erosion and that the joint once extended a considerable distance upward into the overlying strata.

Figure 21 shows plumes on a much smaller scale in massive siltstone. The pattern as a whole is less coarse than that on the joint surface in the previ-

FIG. 20.—Prominent plumose structure on face of systematic joint in massive sandstone of Halgaito.

ous figure. Aside from the difference in scale, however, the patterns are similar.

Figure 22 illustrates that plumes are not invariably parallel with bedding. The joint surface at the left of the hammer shows a plume in which the axis is oriented normal to bedding.

Figure 23 shows in detail the terminal features of a joint surface. The orientation of the ridges of the plume show that the axis is parallel with the bedding and that the left-hand edge of the surface is being observed. At this edge, the axis of the plume is no longer identifiable as it has given place to the radial pattern common at either end. The ridges have diverged to a considerable extent and the pattern has become coarse. The photograph shows the manner in which certain ridges of the plume become more prominent and terminate in small *en échelon* fracture surfaces The writer designates these small fractures "terminal offset faces." They correspond precisely with the "border" or "b-planes" of Woodworth (1897, p. 169). Small-scale plumes are present on the terminal offset faces, a feature noted also by Woodworth (p. 171). Study thus far has shown no systematic relation between the direction of offset of these small surfaces and the larger faces of joints.

Plumose structure is not restricted to systematic joints but is found also on the faces of non-systematic joints as shown in Figure 24. Here the plume is complete, including the terminal offset fractures of the joint edge, but it lacks the oriented axis. No example of plumose structure

FIG. 22.—Plumose structures on faces of systematic joints in massive sandstone of Halgaito. Vertical plume shows on face of joint in left center of photograph. Bedding horizontal.

having a well defined axis was observed on non-systematic joints and this condition is believed restricted to systematic joints.

CONCHOIDAL STRUCTURE

The surfaces of some joints exhibit pronounced conchoidal features. The conchoidal appearance is produced by well defined curvilinear ridges that

FIG. 21.—Plumose structures on faces of systematic joints in massive sandstone of Halgaito. Height of rock about 5 feet.

FIG. 23.—Terminal offset fractures at one edge of systematic joint in massive sandstone of Halgaito. Small plumes on terminal fractures. Surface area shown is about 6 square feet.

FIG. 24.—Plumose structure and prominent terminal offset fractures on face of non-systematic joint in massive sandstone of Halgaito.

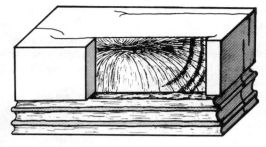

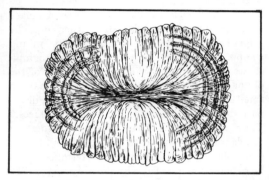

FIG. 25a (Upper).—Schematic block diagram showing plumose structure on face of joint and relation between joint face and joint trace.
25b (Lower).—Diagrammatic sketch of plumose and conchoidal structure on complete face of systematic joint.

occur near the edges of a joint surface. These ridges intersect the faint ridges of the plumose pattern nearly at right angles, and, if several conchoidal ridges are present, they are concentric around the center of the joint surface and appear to have maximum relief along the axis of the plume. The two types of ridges do not interfere visibly with each other but the plumose ridges appear superposed across the larger conchoidal ridges.

Conchoidal ridges are shown clearly on the first two joint surfaces on the left side of Figure 21. The first surface shows a series of concentric ridges at right angles to the ridges of the plume and the second shows only one such ridge on the upper half of the joint face.

RELATION BETWEEN SURFACE FEATURES ON JUXTAPOSED JOINT SURFACES

Observation of joints in several places where the two faces had been separated recently by mechanical action showed that, in each example, the patterns of the opposed faces were complementary in every respect, the ridges of one face corresponding exactly with the troughs of the other. This observation has been made also by Parker (1942, p. 396).

CONCEPT OF JOINT SURFACE

The observation of the surface features on many partly exposed surfaces of systematic joints permits a composite reconstruction of these fea-

tures as they would appear on a completely exposed fresh surface. Figure 25 a-b is such a reconstruction.

The lines of the drawings represent the ridges of the plumose pattern and must be considered diagrammatic. The direction of offset of the terminal fracture faces around the edge of a complete joint surface has not been observed. Inspection of partly exposed surfaces shows that the direction of offset is the same for any joint over the part observed. The mode of offset shown in Figure 25b is conjectural.

Whether or not conchoidal structure commonly appears at both ends of a plumose pattern as shown in Figure 25b is also conjectural.

Inspection of joint surfaces exhibiting a plumose pattern shows that such surfaces tend to be somewhat elongate in the direction of the axis of the pattern. The plumose pattern originates at a point or small area near the center of the surface. The ridges of the plumes have minimum spacing and relief adjacent to this point and along the axis of the plume. These characteristic ridges give the pattern a definite orientation, a condition apparently restricted to systematic joints.

The point of origin of the plumose pattern may lie at the intersection between the joint surface and an identifiable bedding surface but, in many examples, this relation is not found, as in massive, fine-grained sandstones. The point of origin of the plume may lie at some point above or below the bed in which it is observed and the pattern will give a clue whether the edge or center of the joint is being observed.

ABSENCE OF MOVEMENT

With very few exceptions, neither systematic nor non-systematic joints show any evidence of transcurrent movement. Movement normal to joint faces, if present, is microscopic and can be compared with that expected across a crack in a quartz sand grain.

The surface features of joint faces suggest strongly that joints are fractures that begin at a point and propagate outward through the rock. This requires no lateral movement of the rock adjacent to the fracture. Unless there has been subsequent disturbance of a joint and separation of the faces, the joint appears to be a "closed" fracture, that is, the two faces are in contact.

The interlocking nature of the plumose patterns on the opposed faces of a joint also precludes transcurrent movement. No such movement can occur without obliterating the plumose pattern.

CONCLUSIONS—SYSTEMATIC JOINTS

The occurrence of systematic joints as sets of closely and evenly spaced fractures indicates that the force or forces responsible for their formation must have been relatively homogeneous in both intensity and direction.

The spacing of joints commonly varies from set to set in the same area and in the same rock unit. This indicates that the forces causing jointing, although probably homogeneous as shown by the equal spacing of joints within the sets, vary slightly in intensity with respect to the different sets.

Variation in joint spacing from set to set might also indicate anisotropic behavior of the rock in response to a stress although no obvious lithologic anisotropies are present in many of the mapped rock units (i.e., Navajo sandstone).

The nature of the forces that cause jointing is difficult to envision, primarily because of the characteristic lack of discernible displacement along joints. This condition indicates that the stresses in the vicinity of the joint were relieved or dissipated in forming the joint. Otherwise, additional stress or a continuously applied stress should result either in enlargement of the joint or displacement normal to, or parallel with the faces.

A joint is commonly nearly a plane surface in the vicinity of the point of origin and becomes less planar and the plumose pattern diffuse toward the edge of the joint. The terminal offset faces at the fringe of the joint show maximum deviation from the mean strike of the joint face. These features are interpreted here as indications of a lessening in intensity of stress as the fracture propagates outward from the point of origin.

There are, for practical purposes, an infinite number of inhomogeneities in rocks (voids, cracked grains, etc.) that might be expected to serve as points of origin for joints, yet the distribution of systematic joints through the rock is relatively even. From this condition it is inferred that a homogeneous stress was acting on the rock at the time of rupture and that the spacing of the joints reflects the intensity of the stress, being the number of fractures required to dissipate the forces involved.

SIMILARITIES TO FATIGUE AND STATIC FRACTURE

Oriented plumose structures and conchoidal markings like those of systematic joints are found on the faces of fatigue and static fractures in metals and concrete. Such fractures have been observed both experimentally and in the field and theoretical explanations advanced to account for them. Experimental work done thus far has been concerned primarily with metals, and the observed fractures are formed under stress conditions not thoroughly understood. The similarity between these fractures and joints is so striking, however, that a direct comparison between the two seems useful even though the extent to which they are analogous is unknown.

Both fatigue and static fractures are initiated at some point (ordinarily where there is an observable inhomogeneity) and are propagated outward through the material, commonly as a planar fracture. In a thin plate the fracture will be nearly perpendicular to the upper and lower surfaces of the plate. The manner in which such fractures may be initiated and the forces directly responsible for the growth of the fracture have been treated both theoretically and experimentally by Griffith (1924) and are reviewed briefly in the following discussion.

Attempts to calculate the stress at which a brittle material should fail in pure tension invariably give higher values than those obtained in the testing. Such behavior (Griffith, 1924) results from minute cracks in the tested material. These cracks are primarily surface phenomena on glass; in other materials they may be distributed throughout with random orientations. With the application of a force, tensile or otherwise, certain of the cracks, "dangerous cracks" so called because of their greater size compared with other cracks or because of favorable orientation with respect to the imposed force, will be enlarged as a result of the abnormally high stress at their edges. Rupture will eventually take place provided the stress value is high enough. Such cracks have also been observed to form spontaneously in steel and glass. This is known as static fatigue and the mechanism involved is not completely understood.

The theoretical explanation for the concentration of stress at the edges of such a small crack is reviewed concisely by Jaeger (1956, p. 85) as follows.

"The effect of a crack is to produce a very high concentration of stress at its edge. The amount of this can be calculated from the result that the maximum tensile stress in a flat plate containing an elliptical hole of major axis 2 l and subjected to an average tensile stress in a direction perpendicular to the major axis is given by

$$2\left(\frac{l}{p}\right)\frac{1}{2} \qquad (1)$$

where p is the radius of curvature (of the crack) at the ends of the axis, and as $p \rightarrow \phi$, that is, the ellipse tends to a flat crack, the stress tends to infinity . . ."

Orowan (1949, p. 194) prefers to disregard the radius of curvature of the crack where it is of an order of magnitude less than the intermolecular distance between adjacent molecules. The result of the calculations is not significantly different from that obtained considering an ellipsoidal crack. Griffith assumed that such minute cracks would enlarge if the work of the external forces is enough to cover the increase in surface energy

when there is a slight increase in the length of the crack. The crack then becomes unstable with respect to the imposed stress and begins to run.

Orowan (1949, p. 202) states that the strength of a polycrystalline metal increases with decrease in grain size and pictures a fracture as beginning in one grain and spreading to adjacent grains. ". . . The first question is whether the initiation of a crack in a grain, or its propagation to the next grain, needs a higher stress. There is no certain answer to this question, but apparently the propagation of a crack across grain boundaries is, in general, more difficult than its initiation. There are two possibilities for the extension of a crack from one grain to another. Either it may directly penetrate the grain boundary; or it may produce in the neighboring grain a high stress which starts there an incipient crack. In either case, the condition for the continuation of the propagation across the boundary is probably the attainment of a critical stress at the end of the crack (i.e., at the boundary), or within a certain region of the next grain . . ."

Whether the mechanism of crack initiation and propagation in polycrystalline metals and in sedimentary rocks is entirely similar is not known certainly, although observation of fractures in both concrete and metal suggests that the mechanism is similar. The propagation of a crack depends primarily on the abnormal concentration of stress at its leading edges and, in this respect, the requirements for crack propagation should be similar in any brittle material. In sedimentary rocks, which are aggregates of discrete grains closely or loosely packed and bonded more or less firmly by some cement, there are theoretically and actually many small flaws such as pore spaces and fractured grains. There are then innumerable points within any rock unit where a fracture could start under a variety of stress conditions. Even though such fractures may form under a variety of stress conditions, they have the common feature of starting at some point and being propagated outward through the material.

Fatigue fracture is caused by cyclic stresses such as repeated loading and unloading and certain empirical concepts concerning this type of failure have been worked out from observations on specimens of metals and concrete. Empirically a brittle material breaks in fatigue after a certain number of applications of a stress less than that required to break the material under static conditions. There is, nevertheless, in some cases at least, a critical stress value below which the material will not fracture regardless of the number of stress cycles. Stress values below the critical amplitude are called safe and values above, unsafe. The critical amplitude in any brittle material is very nearly the same for repeated loading and for alternating positive and negative loads.

Orowan (1949) states that several different mechanisms may lead to the formation and propagation of fatigue cracks and that the fracture may be trans-crystalline or inter-granular. Thus the main features of fatigue fracture depend primarily on the presence of structural inhomogeneities and on the capacity for strain hardening in the material rather than on a unique type of stress condition.

Orowan notes as a significant fact that fatigue phenomena are not affected significantly by the frequency of the stress cycles. His theoretical interpretation of the stress conditions leading to fatigue fracture is based on the behavior of a "plastic" body of small size within an elastic body. The plastic body could be the result of stress concentration around a small crack, a crystal, or grain with favorably oriented slip surfaces, or any other inhomogeneity.

Orowan states (1949, p. 224) ". . . The salient point about inhomogeneities is that if a mainly elastic body is subjected to cycles of constant stress amplitude, the plastic strain amplitude in small strain-hardening plastic inclusions in the body is not constant but diminishes in the course of the cycles. According to the fundamental theorem of fatigue (Orowan, 1939), proved below, the total plastic strain in such an inclusion always converges toward a finite value unless the stress-strain curve becomes horizontal. This value increases with the amplitude of the stress applied to the body, and the amplitude is safe or unsafe according to whether the limiting total plastic strain in the inclusion is above or below the critical strain for fracture (if fracture obeys a stress criterion, the critical stress corresponds to a critical strain according to the stress-strain curve)."

Under this theory a crack would spread until it reaches a length where at the next application of the full *tensile* stress of the cycle a complete fracture is produced. Fatigue thus offers a mechanism for producing fractures at stress values well below the static breaking strength of brittle materials.

It can not be said, on the basis of the data

presented thus far, whether joints belonging to different sets in an area are formed in response to the same forces or whether they are even of the same age. It is not possible to infer the sense of the forces causing the jointing on the basis of individual local observations because the angular relations between sets of joints are not consistent and do not appear to follow any discernible rule. Their characteristic disposition indicates only formation in a relatively homogeneous stress field by forces dissipated in forming the joints. The manner in which a systematic joint forms, however, is probably such that it can be propagated across another systematic joint (provided that joint is closed) without deflection or without displacement of the first joint. If the joints of one set can be propagated across the joints of another set without deflection or displacement then the problem of relative ages becomes enigmatic.

The spatial relations of joints give some qualitative evidence of the time of formation. The relations between joints and between joints and structure rock units are similar regardless of the disposition of the rock units. This indicates either that the joints were formed before the strata were tilted or that they were formed by some force not influenced by the dip of the strata. A zone of steeply tilted strata (in an area of relatively flat-lying strata) should result in an inhomogeneity with respect to any imposed regional stress. The joints do not show the effect of such a structural inhomogeneity, because the distortion of the regional pattern across tilted strata is purely geometric. This would imply that the formation of the joints preceded tilting of the strata.

CONCLUSIONS—NON-SYSTEMATIC JOINTS

The physical characteristics of non-systematic joints, such as irregular or sharply curved traces, lack of oriented surface structure, and termination of the joints at bedding surfaces and against each other, indicate formation under non-homogeneous stresses.

The stress fields are probably local and of a low order of magnitude inasmuch as non-systematic joints are not propagated across each other or across systematic joints.

They appear to be related directly to the behavior of the relatively small block in which they occur and so would be secondary structures occurring after the formation of the systematic joints.

Where a rock has been cut by a single set of

systematic joints, it consists essentially of a series of flat slabs of rock. If several sets of joints are present, the rock is cut into polyhedral blocks of different shapes depending on the angle of intersection of the joints.

Any warping or bending of the strata subsequent to the formation of the systematic joints should generate stresses and, consequently, fractures in the blocks between these joints. The common occurrence of a right-angle or nearly right-angle juncture between the traces of non-systematic and systematic joints indicates some control exerted by the systematic joint on the forces producing the non-systematic joint.

It is probable that, whatever the distribution of stresses, all non-systematic joints in a block form at very nearly the same time because the terminations of these joints against each other are very definite.

REGIONAL RELATIONS

The regional relations of systematic joints with other structures are expressed primarily in terms of the fracture pattern produced by the disposition of the joints in plan. Individual joint trends, or the pattern as a whole, can then be compared directly with the trends of other structural elements such as folds, faults, and dikes.

Rather than use the more conventional representations of joints (such as rosettes of dial diagrams), the writer represents the major elements of the regional joint pattern of the Comb Ridge-Navajo Mountain region by an arbitrary network of fine lines, evenly spaced in such a manner that the areal extent of the different joint trends can be shown directly and so that the other features of the map are not obscured (Pl. I). Blank areas in the joint pattern are areas where sand-cover prevents observation of joints both in the field and on aerial photographs.

REGIONAL JOINT PATTERN

A complex, overlapping series of joint trends forms the regional pattern. Each trend appears to have vague and ill-defined areal limits and all trends mapped extend for an undefined distance in some direction beyond the mapped area.

Several trends appear to be broadly arcuate over great distances. Deviations from a straight line are rarely more than 5°, however, and this deviation is not enough to determine unequivo-cally whether angular relations between trends change significantly.

In addition, trends may show slight local deviations caused by geometric distortion along flanks of folds where the dip of the strata is steep (30° or more). The value of the dip of the strata and the angle at which the trend intersects the axis of the fold determines the extent of the deviation.

ANGULAR RELATIONS BETWEEN TRENDS

The angle of intersection between two joint trends remains essentially constant over large areas. The value of the angle of intersection, however, ranges from less than 15° to 90°.

The various trends can be considered as individual units or as related groups of two or more trends intersecting at various angles, according to the prejudices of the observer. The elements of the joint pattern west of the Organ Rock monocline, for example, could be arranged in several different groupings with sets intersecting at angles of about 60°, 90°, or even 120°. Such neat groupings within the pattern are not everywhere possible, however, as in the area east of the Comb monocline where trends are present that intersect at an angle of about 15°.

Intersecting trends appear to have no effect on each other and even as individual systematic joints, they cross each other without deviation. Angular relations between trends appear to follow no rule other than that they are essentially constant over large areas. In addition, a trend can terminate in any direction independently of the other trends.

ANGULAR RELATIONS BETWEEN JOINT TRENDS AND FOLD AXES

The most impressive feature of the regional joint pattern is that it does not appear to be affected materially by the presence of folds. Each trend or element of the regional pattern crosses several folds without significant deviation and does not swing to maintain an angular relation to a fold axis as the axis changes direction. For example, the east-west joint trend crosses several anticlinal axes at right angles in the western part of the region but in the east crosses the Capitan anticline at angles ranging from about 45° to 70°. The only obvious correlation between fold axes and joint trends is that the fold axes tend to lie more or less parallel with some fracture direction.

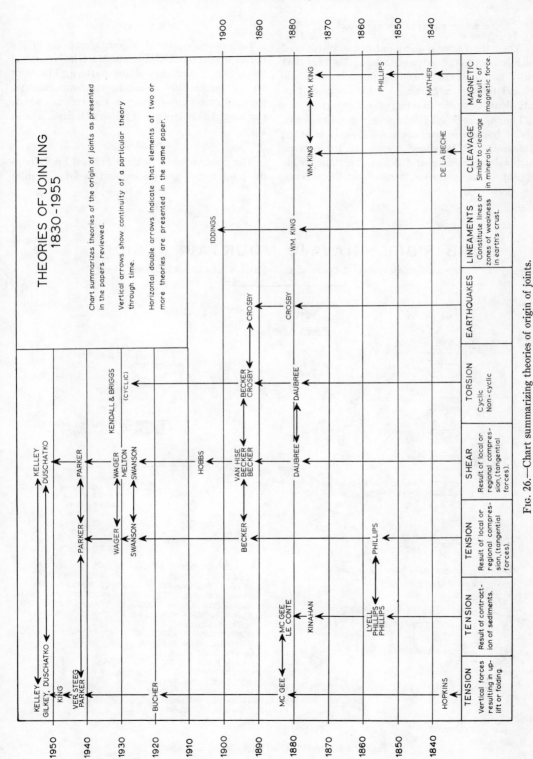

Fɪɢ. 26.—Chart summarizing theories of origin of joints.

JOINT PATTERN IN SECTION

When the regional joint pattern is considered with respect to the outcrop pattern, it is clear that individual joint trends and the pattern as a whole extend through a geologic section ranging from late Pennsylvanian to late Jurassic in age.

Thus the regional joint pattern persists across period boundaries and the era boundary between Paleozoic and Mesozoic. Details of the pattern such as the presence or absence of certain trends appear to be consistent throughout the section.

AGE RELATIONS

Just as with individual joints, nothing in the regional joint pattern indicates the relative ages of the different elements of the pattern. The joint trends conceivably could be contemporaneous, penecontemporaneous or, as the trends can be considered as independent units, of quite different ages.

CONCLUSIONS

The constant direction maintained by individual joint trends and consequent variable angular

GEOLOGIC MAP

OF THE

COMB RIDGE – NAVAJO MOUNTAIN REGION

COCONINO, NAVAJO AND APACHE COUNTIES, ARIZONA - SAN JUAN COUNTY, UTAH

SCALE-MILES

ROBERT A. HODGSON

1960

EXPLANATION

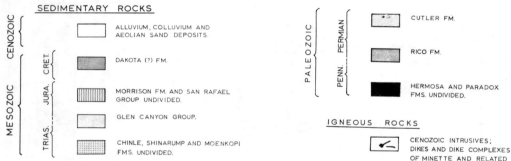

SEDIMENTARY ROCKS

CENOZOIC
— ALLUVIUM, COLLUVIUM AND AEOLIAN SAND DEPOSITS.

MESOZOIC
JURA. CRET.
— DAKOTA (?) FM.
— MORRISON FM. AND SAN RAFAEL GROUP UNDIVIDED.
— GLEN CANYON GROUP.
TRIAS.
— CHINLE, SHINARUMP AND MOENKOPI FMS. UNDIVIDED.

PALEOZOIC
PENN. PERMIAN
— CUTLER FM.
— RICO FM.
— HERMOSA AND PARADOX FMS. UNDIVIDED.

IGNEOUS ROCKS

— CENOZOIC INTRUSIVES; DIKES AND DIKE COMPLEXES OF MINETTE AND RELATED BASALTIC ROCKS.

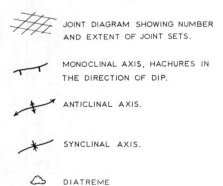

MAP SYMBOLS

— JOINT DIAGRAM SHOWING NUMBER AND EXTENT OF JOINT SETS.

— MONOCLINAL AXIS, HACHURES IN THE DIRECTION OF DIP.

— ANTICLINAL AXIS.

— SYNCLINAL AXIS.

— DIATREME

INDEX

1. NAVAJO MTN. DOME
2. TUBA CITY SYNCLINE
3. BEAVER CREEK ANTICLINE
4. RAPID SYNCLINE
5. BALANCED ROCK ANTICLINE
6. BALANCED ROCK MONOCLINE
7. NOKAI SYNCLINE
8. COPPER CANYON ANTICLINE
9. COPPER CANYON SYNCLINE
10. ORGAN ROCK ANTICLINE
11. ORGAN ROCK MONOCLINE
12. OLJETO SYNCLINE
13. DOUGLAS MESA ANTICLINE
14. MITTEN BUTTE SYNCLINE
15. HALGAITO ANTICLINE
16. CAPITAN ANTICLINE
17. MEXICAN HAT SYNCLINE
18. RAPLEE ANTICLINE
19. COMB MONOCLINE
20. TYENDE SYNCLINE
21. WATER CREEK ANTICLINE
22. DEFIANCE ANTICLINE
23. CHINLE SYNCLINE

Legend for map (Pl. I) on pages 30–31 and 34–35.

relations with folds suggest that the joints precede the folding in this region. On the other hand, the angular relations between joint trends and fold axes suggest that the folds were influenced by, and followed, one or more joint trends. The constant direction of individual trends for great distances also indicates that the joints were formed by homogeneous forces acting over a large area.

Origin of Joints

Though numerous fundamentally different hypotheses have been advanced to account for systematic and non-systematic joints, no hypothesis is universally accepted at the present time. Figure 26 shows, in graphic form, several of the major hypotheses advanced to account for joints and the sequence in which they appeared in the literature. No attempt at an extensive review of these papers is made here (see Hodgson, 1958).

These hypotheses (except those attributing joints to magnetic forces or forces of crystallization) consider joints as shear or tension fractures resulting from regional or local compression, tension, torsion, or some combination of these forces. At present, jointing is generally considered as genetically related to folding both in time and space (Billings, 1954; de Sitter, 1956; Gilkey, 1953; Goguel, 1952; Nevin, 1949). For example, where a fold is believed to be compressional, two sets of joints are considered as compression or shear joints if they intersect at an acute angle facing the direction of the compressive force as inferred from the fold. In contrast, if the joints intersect at nearly right angles, one set downdip and the other along the strike of the fold, they are interpreted as tension joints.

The map showing the regional joint pattern (Pl. I) of the Comb Ridge-Navajo Mountain area indicates that, in plan, the fold axes are commonly arcuate or sigmoidal, but the joint sets maintain a relatively constant direction.

Thus, under prevailing theory, the same joint set might be considered tensional in one area and compressional in another on the basis of the angular relations between the joint set and the fold axes. The ideal theoretical condition where only *two joint sets* are present and cross a fold at such an angle that they can be interpreted as tensional or compressional is not found in the Comb Ridge-Navajo Mountain area. Also, the individual joints show no obvious features that

would differentiate "tension" from "compression" or "shear" joints. For these reasons, any theory that postulates that systematic joints are genetically related to folding is rejected for this region. Whether regional compressional or tensional forces are directly responsible for the jointing must be decided on other criteria.

Age of Systematic Joints

Prevailing theories of jointing could be better evaluated if the time of formation of systematic joints was known relative to the time of consolidation of the rocks and to time of folding. Evidence, direct and indirect, suggests that joints form very early, at least in sedimentary rocks.

Kinahan (1875) describes jointing in both "recent" and "older" stratified deposits (recent deposits being poorly consolidated or unconsolidated sediments). The joints are classified as "master" and "minor." Minor joints are limited to a single bed and are irregular whereas master joints are regular, parallel, and cut through many beds. His classification of joints holds true for both recent and older rocks.

Gilbert (1882a) states specifically that joints in clays of the former bed of Great Salt Lake form sets that intersect nearly at right angles and control the local drainage pattern. The "principal" series of joints strikes north-south and the "subordinate" series east-west. The thin laminae of the lake sediments show no displacement across either series of joints.

Crosby (1885) describes systematic joints in Miocene clays and sands of the Nomini cliffs on the west shore of the Potomac River, Westmoreland County, Virginia. One set is nearly parallel with the cliff face and the other set at about right angles to it. The clays are described as "wet."

Kendall and Briggs (1933) consider cleat in coal and systematic jointing as directly comparable. They state that cleat is found in coal ranging in rank from lignite to bituminous. It is best expressed in bright parts of bituminous coal but has been found also in lignite where joints have not yet formed in juxtaposed sands and clays

These observations show that systematic joints form in young and poorly consolidated sediments. No such sediments have wide areal distribution in the Comb Ridge-Navajo Mountain area and it is not possible to date the joints in other rocks directly except within very wide limits. The minette dikes of the area are intruded along sys-

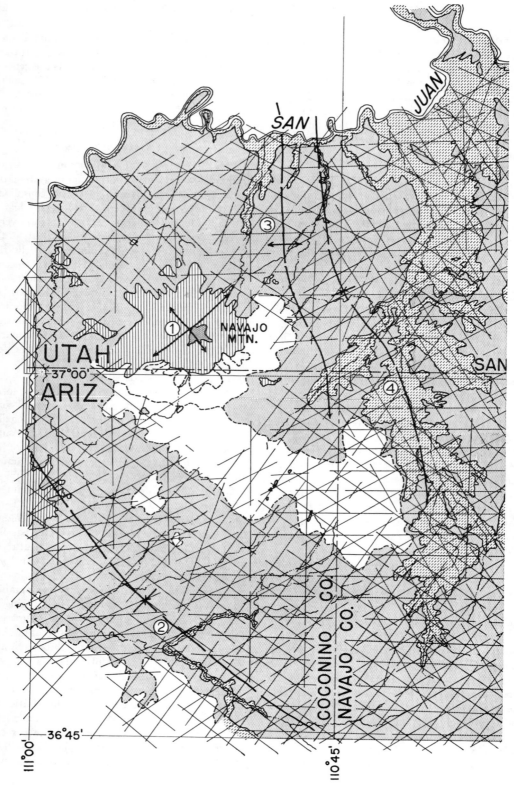

Pl. I, Pt. 1.—Comb Ridge-Navajo Mountain region.

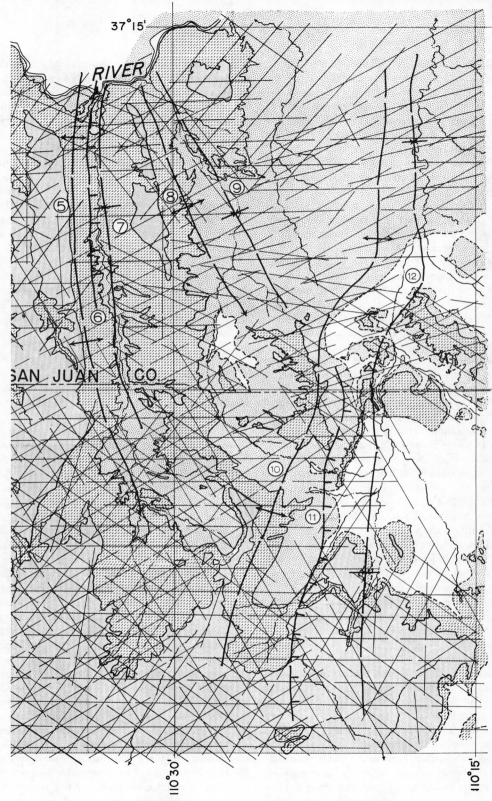

Pl. I, Pt. 2.—Comb Ridge-Navajo Mountain region.

tematic joints in exposed Permian and Jurassic rocks. These dikes are dated as "Tertiary" by Gregory (1917) and Baker (1936); erosion of the Cenozoic strata prevents more precise dating. Joints intruded by these dikes are older than the dikes and are penecontemporaneous with, or younger than, the strata in which they occur.

The spatial relations between joints and the rocks in which they occur are similar regardless of the dip of the strata. Distortion of a joint pattern locally along the steep flank of a fold is then, purely geometric, a matter of rotation. From this it is inferred that the jointing precedes the folding which in this area is "Laramide."

A possible clue about the age of the jointing may be found on the basis of indirect evidence such as the disposition of the joints with respect to the surfaces of structure rock units and bedding.

With few exceptions, systematic joints are nearly perpendicular to the upper and lower surfaces of flat-bedded strata *regardless of the dip of the strata*. Such a relation can be explained if the bedding itself exerts some influence on the forces involved so that joints are invariably produced normal to the bedding surfaces. Thus the influence of bedding must outweigh all other influences and be the controlling factor in the orientation of the joints. It seems improbable that such uniformity of disposition of the joints should be obtained otherwise in areas of folding where the structural geometry and the stress conditions are complex.

If the bedding surfaces exert a strong influence on the disposition of joints in flat-bedded rocks, they should do the same in the prominent cross-bedded Permian and Jurassic sandstones of the area. One would expect to find that joints tended to be normal to the cross-bedding. The original dip of the cross-beds ranges from about 20° to 25°. With few exceptions, however, the systematic joints of these sandstones are nearly perpendicular to the upper and lower surfaces of the gross rock unit (Figs. 6 and 15) and not perpendicular to the cross-beds.

Apparently, however, the orientation of bedding has little influence on the dip of systematic joints. The joints are perpendicular to originally flat-lying beds but not to beds having a significant original dip. The inference drawn here is that the joints are formed essentially normal to a tangent to the earth's surface, early in the history of a sediment and before any significant tilting or warping of the strata has occurred. Such reasoning implies

that the joints are formed successively in each newly deposited rock unit, possibly as soon as the rock is capable of being fractured. If this be true, then each newly deposited rock unit, upon lithification, may be considered, to scale, as an extremely thin, brittle sheet. Fractures occurring in such a material would be more or less normal to the upper and lower surfaces of the sheet. The Navajo sandstone, for example, can be considered as such a thin sheet. The systematic joints are normal to the upper and lower surfaces of the gross sandstone unit but not to the individual cross-beds. The gross lithologic unit, therefore, is also a structure rock unit with respect to the jointing.

Before considering further these limiting conditions, it is worthwhile to review briefly the several theories of fracture that might be applied to explain systematic jointing in the mapped area on the basis of the regional pattern alone.

SHEAR THEORY

The regional joint pattern can not be used as the basis for explaining joints by shear without making equivocal assumptions concerning the genetic relations between joint sets.

The shear theory states that failure will occur at some point where the maximum shear stress exceeds the shear strength of the material, and hence that the planes of fracture will bisect the angle between the greatest and least principal stress axis and include the intermediate axis. In this form the theory predicts that the strength of the material will be equal in both compression and tension, a condition not confirmed by experiment. Fractures actually form along planes at some angle less than 45° to the direction of the maximum principal stress. When the coefficient of internal friction is introduced (constant for a particular material), theoretical and experimental results agree that the acute angle between two sets of fractures will face the direction of the maximum principal stress.

The joint pattern of the Comb Ridge-Navajo Mountain area shows that essentially constant angular relations are found in plan. It is assumed that the direction of the intermediate principal stress is essentially vertical and the maximum and minimum principal stresses horizontal. The shear theory requires that only two joint sets at one time be considered as congeneric. Assignment of a genetic relation between any two sets where more than two sets are present must be arbitrary and this presents a dilemma. In the western part of the mapped area, for example, there are in places six well defined joint sets. These six sets may be divided into three groups of two sets each in more than one way. They may be grouped so that any two sets will intersect at about 90° or so that they will intersect at about 30°. (The value in the field may differ by as much as 5° from the average value.) If the angle of intersection is close to 90°, it is difficult to determine unequivocally in which of two directions lies the maximum principal stress. In addition, under the shear theory such an angle would indicate a coefficient of friction close to zero. If the angle of intersection is close to 30° there is no difficulty in inferring the direction of the maximum principal stress. In either case, however, we would be compelled to infer that the several different pairs of sets were formed at different times by compressive forces that would, at a minimum, have to operate from directions separated by nearly 90°.

In other parts of the mapped area we find that the joint sets can not be so conveniently classed into congeneric groups of two and that the angles of intersection between any two sets may range from 15° to 65°. Thus it becomes difficult or impossible to make any reasonable decision as to which sets are genetically related, particularly if there is an uneven number present. Where the angle of intersection is very low (15°) the value of the coefficient of friction must be high. There is, however, nothing in the nature of the joints or in the nature of the rocks to indicate any great change in the coefficient of friction from area to area as would have to be postulated on the basis of the changes in angles of intersection.

Individual joints do not show evidence that they are shear fractures. There are no slickensides or other features that would indicate a tendency for the faces of the joints to move past each other.

TENSION THEORY

According to the tension theory, a brittle material should fracture along a plane at right angles to the direction of maximum tensile stress. Joints show surface features compatible with the idea that they have formed under tension. Explanation of the regional joint pattern by tension, however, is subject to complications similar to those if the shear theory is assumed. Each joint set can be considered as an individual unit formed under a tension acting normal to the strike of the set. The forces must act, therefore, in a different direction to produce each set.

TORSION

According to the torsion theory, two sets of fractures should form nearly at right angles. Thus, the same difficulty of choosing appropriate sets of congeneric joints and explaining the angular relations between sets applies to this theory as to the shear theory.

ALTERNATIVE HYPOTHESIS

If joints form early in the history of a sediment, as observed directly in a few examples and inferred on indirect evidence for others, then systematic joints must be successively younger upward through the section. The joint pattern would be imposed on each new sediment at some later time when it is capable of fracture and before significant warping or tilting has taken place.

The difficulties involved in explaining the regional joint pattern by simple shear or tension would apply here but would be compounded because the same distributions of stresses would be required to operate successively in the same directions over a large segment of geologic time to account for the presence of the complex joint pattern through the section.

Some other mechanism must be operative if regional compression or tension does not directly cause joints. Fatigue, resulting from cyclic stresses, is a possible mechanism. The fractures formed under fatigue show features similar to those of joints and, in the case of recently deposited sediments, the mechanism would operate under optimum conditions (a thin sheet of unfractured rock under minimal confining stress).

Kendall and Briggs (1933) have treated this problem, suggesting that semi-diurnal earth tides produce joints by a torsion mechanism (as suggested by Daubree, 1879) that controls the direction of the joints as well as their formation. This concept is based on the belief that the two chief directions of jointing in the northern hemisphere are generally northeast and northwest and so coincide with the direction of maximum shear stress caused by tidal forces acting nearly parallel with the equator. This assumption concerning the chief directions of jointing is not warranted for the Comb Ridge-Navajo Mountain area and probably is not valid for many other regions.

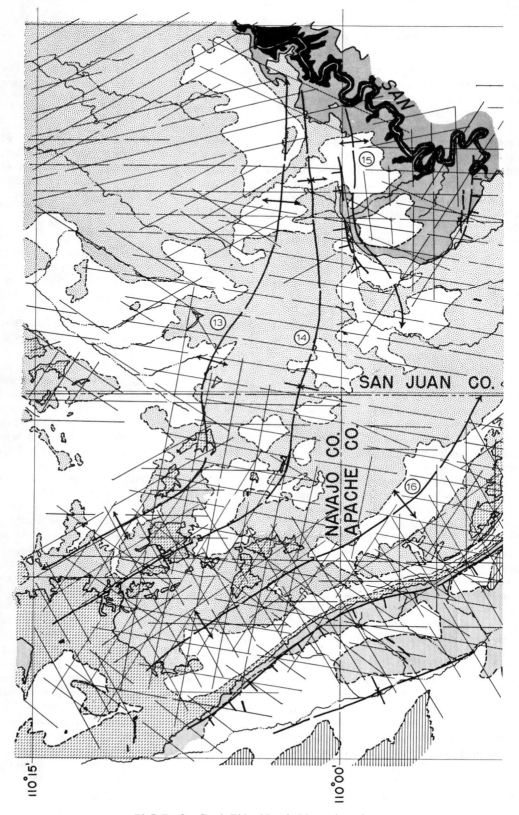

Pl. I, Pt. 3.—Comb Ridge-Navajo Mountain region.

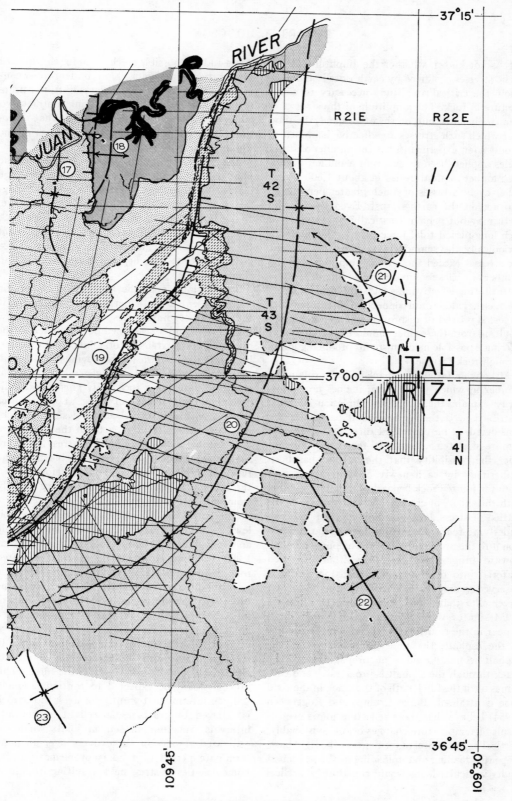

Pl. I, Pt. 4.—Comb Ridge-Navajo Mountain region.

It is not known whether the amplitude of the cyclic stresses produced by earth tides is above or below the critical amplitude necessary to cause fracture in rock. The amplitude of the stresses is admittedly small but, on the other hand, the time over which such stresses are effective is immense. In only a few experiments has the number of stress cycles applied been in excess of a million cycles, the number of tidal cycles in about 1,500 years. Other cyclic stresses of much greater periodicity also act on the earth's crust. Even though the earth may not behave elastically in response to such long-period tidal forces, they may reinforce periodically the semi-diurnal earth tides so that the stresses exceed the critical amplitude for the number of cycles required to cause fracture.

Until the effectiveness of semi-diurnal earth tides, either alone or in connection with other tidal forces, is proved or disproved, the cyclic stresses resulting from these tides should be considered as offering a possible mechanism for producing joints through rock fatigue.

Although a fatigue mechanism may explain the manner in which joints form, it does not necessarily explain their constant direction or areal distribution. If the regional joint pattern is not considered to be controlled directly by regional compression, tension, or tidal forces, we can consider the possibility that the pattern has been "inherited" by each newly deposited rock unit from the jointed rock beneath, a possibility in line with the theory of crack propagation, the concentration of stress at the leading edges of joints in underlying rocks permitting their upward extension into the unfractured rock. The phenomena of upward propagation of cracks from fractured material into overlying unfractured material is observed commonly in concrete or other rigid base pavements that have been resurfaced. Roberts (1954) states that the time over which cracks appear in the resurfacing depends on the rigidity of the asphalt, the time being less where the asphalt is more rigid. The propagation of the crack through the asphalt is progressive and continues until the full length of the crack in the rigid base is attained. Bone, Crump, and Roggeveen (1954) believe that these reflection joints are the result of temperature changes or, more probably, vertical or horizontal movements in the rigid base. Repeated loading and unloading of the pavement under vehicular loads would constitute a cyclical

stress and is a possible mechanism for the upward propagation of the crack. Thus, under proper conditions, fractures can be "reflected" from a fractured material into an overlying unfractured material. In this respect it is possible that reflection joints are analogous with the joints of the Comb Ridge-Navajo Mountain area.

The directions of the different sets that form the joint pattern in the mapped area do not change perceptibly through a thick section of rocks ranging from Pennsylvanian through Jurassic in age. If the joints form successively in each new rock unit, there must be either the upward reflection of the pattern and (or) a remarkably consistent disposition of the stress fields of the causal forces in time and space.

The variations in the spacing of systematic joints present a particularly knotty problem. If the spacing of joints varies from bed to bed, how can all the joints in each bed be considered as the direct upward extension of joints in the underlying rock? Within the limits previously discussed, the spacing of the joints is in general proportional to the thickness of a rock unit. This would indicate that the intensity of stress at the time of jointing is similar in rock units of similar thickness, a condition to be expected if the rocks were *at or near the earth's surface.* If the stress field were slightly inhomogeneous (uniaxial or biaxial as the result of contemporaneous tectonic deformation), the spacing might be expected to vary slightly from set to set as observed.

The entire process would be visualized as follows. When a thin layer (considered here to scale and so including structure rock units as much as 1,000 feet thick) of newly deposited, nonjointed rock is at the point of rupture throughout as a result of fatigue engendered by tidal forces, the extension of a joint in the underlying rock upward into the unfractured rock should relieve stress adjacent to the extended joint. This in turn should produce a sudden tension at right angles to the joint in the surrounding area and so initiate additional joints parallel with the first. This process would be expected to continue sequentially or intermittently until the stress in the rock is relieved by fractures spaced at appropriate intervals. Whether all sets of joints would be generated at once or successively over an indeterminate period of time is problematical. Inhomogeneities in stress fields resulting from local or regional tectonic activity conceivably exert some influence on the direction which will be reflected initially into the unjointed rock unit.

SUMMARY AND CONCLUSIONS

The main conclusion drawn from the present study in the Comb Ridge-Navajo Mountain area is that the systematic joints are not related genetically to the Laramide folding. Prevailing theories concerning the origin of systematic joints and regional joint pattern are rejected on the basis that they can be applied only if a number of assumptions are made for which adequate criteria are lacking.

An alternative hypothesis is offered in which systematic joints are produced by tidal forces through a fatigue mechanism and the direction of the joints is considered to be inherited by upward reflection of the joint pattern in pre-existing jointed rocks. The entire process is thought to occur early in the history of a sediment. This hypothesis is compatible with field evidence and is possible on the basis of present theory. The data to treat the problem quantitatively are lacking, however, and so it must remain a working hypothesis at present. In postulating such a series of events the question of the ultimate origin of the joint pattern is not answered.

In the final analysis, no theory thus far presented in the literature to account for systematic jointing appears to be completely satisfactory. Phillips (1855, p. 42) wrote: "Now, surely nothing can be more certain than the inference that some general and long continued agency, pervading at once the whole mass of these dissimilar and successively deposited strata, was concerned in producing this remarkable constancy of direction in the fissures which divide them all." Additional crucial data are needed before this "general and long continued agency" is defined clearly.

REFERENCES

AMERICAN GEOLOGICAL INSTITUTE, 1957, "Glossary of Geology and Related Sciences," *Natl. Acad. Sci.-Natl. Research Council Pub. 501*, p. 325.

BAKER, A. A., 1936, "Geology of the Monument Valley-Navajo Mountain Region, San Juan County, Utah," *U. S. Geol. Survey Bull. 865*.

———, DANE, C. H., AND REESIDE, J. B., JR., 1936, "Correlation of the Jurassic Formations of Parts of Utah, Arizona, New Mexico, and Colorado," *ibid., Prof. Paper 183*.

———, ———, ———, 1947, "Revised Correlation of Jurassic Formations of Parts of Utah, Arizona, New Mexico and Colorado," *Bull. Amer. Assoc. Petrol. Geol.*, Vol. 31, pp. 1664–68.

———, AND REESIDE, J. B., JR., 1929, "Correlation of the Permian of Southern Utah, Northern Arizona, Northwestern New Mexico and Southwestern Colorado," *ibid.*, Vol. 13, pp. 1413–48.

BASS, N. W., 1944, "Correlation of Basal Permian and Older Rocks in Southwestern Colorado, Northwestern New Mexico, Northeastern Arizona, and Southeastern Utah," *U. S. Geol. Survey Prelim. Chart 7*, Oil and Gas Inv. Ser.

BECKER, G. F., 1893, "Finite Homogeneous Strain, Flow and Rupture of Rocks," *Bull. Geol. Soc. America*, Vol. 4, pp. 13–90.

———, 1894, "The Torsional Theory of Joints," *Trans. Amer. Inst. Min. Eng.*, Virginia Beach Meeting (Reprint). 8 pp.

BILLINGS, M. P., 1954, *Structural Geology*, pp. 106–23. Prentice-Hall, Inc., New York.

BONE, CRUMP, ROGGEVEEN, 1954, "Control of Reflection Cracking in Bituminous Resurfacing Over Old Cement-Concrete Pavements," *Natl. Research Council*, Highway Research Board, Proc. 33d Ann. Meeting, pp. 345–54.

BUCHER, WALTER H., 1920–21, "The Mechanical Interpretation of Joints," *Jour. Geol.*, Vol. 28, pp. 707–30; and Vol. 29, pp. 1–28.

COOPER, JACK C., 1955, "Cambrian, Devonian and Mississippian Rocks of the Four Corners Area," *Four Corners Geol. Soc. Guidebook*, pp. 59–65.

CROSBY, W. O., 1882, "On the Classification and Origin of Jointed Structures," *Proc. Boston Soc. Natural Hist.*, Vol. XXII, pp. 72–85.

———, 1893, "The Origin of Parallel and Intersecting Joints," *Technology Quarterly*, Vol. IV, pp. 230–36.

DAUBREE, A., 1879, *Geologie Experimentale*. Librare des Corps des Points et Chaussees, des Mines et des Telegraphes, Paris.

DE LA BECHE, HENRY, 1839, *Geological Report on Cornwall*. The Geol. Obs., London.

DE SITTER, L. U., 1956, *Structural Geology*. 552 pp. McGraw-Hill Book Company, Inc., New York.

DUSCHATKO, ROBERT W., 1953, "Fracture Studies in the Lucero Uplift," *Tech. Inf. Bull. RME-3072*, Atomic Energy Commission Tech. Inf. Service, Oak Ridge, Tennessee.

GILBERT, G. K., 1882 a, "Post-Glacial Joints," *Amer. Jour. Sci.*, 3d Ser., Vol. 23, pp. 25–27.

———, 1882b, "On the Origin of Jointed Structure," *ibid.*, Vol. 24, pp. 50–53.

GILKEY, ARTHUR K., 1953, "Fracture Pattern of the Zuni Uplift," *Tech. Inf. Bull. RME-3050*, Atomic Energy Commission Tech. Inf. Service, Oak Ridge, Tennessee.

GOGUEL, JEAN, 1952, *Traite de Tectonique*, pp. 52–54. Masson et Cie., Paris.

GREGORY, H. E., 1917, "Geology of the Navajo Country," *U. S. Geol. Survey Prof. Paper 93*.

GRIFFITH, A. A., 1924, "Theory of Brittle Strength," *First Int. Cong. Applied Mechanics*, Vol. 55.

HARSHBARGER, J. W., REPENNING, C. A., AND IRWIN, J. M., 1957, "Stratigraphy of the Uppermost Triassic and Jurassic Rocks of the Navajo Country," *U. S. Geol. Survey Prof. Paper 291*.

HOBBS, W. H., 1911, "Repeating Patterns in the Relief and in the Structure of the Land," *Bull. Geol. Soc. America*, Vol. 22, pp. 123–76.

HODGSON, R. A., 1956, "Preliminary Report on Jointing in the Comb Ridge-Monument Valley Region, Arizona and Utah." (unpublished manuscript).

———, 1958, "A Regional Study of Jointing in the Comb Ridge-Navajo Mountain Area, Arizona and Utah," Ph.D. dissertation (unpublished), Yale University.

HOPKINS, E., 1834, "Phenomena of the Weald and the Bas Boullonnais," Geol. Trans., Vol. 7, p. 598.

HUNT, C. B., et al., 1953, "Geology and Geography of the Henry Mountains Region, Utah," U. S. Geol. Survey Prof. Paper 228.

IDDINGS, J. P., 1904, "A Fracture Valley System," Jour. Geol., Vol. 12, pp. 94–105.

JAEGER, J. C., 1956, Elasticity, Fracture and Flow, p. 152. London, Methuen and Company, Ltd.

KELLEY, VINCENT C., 1955, "Regional Tectonics of the Colorado Plateau and Relationship to the Origin and Distribution of Uranium," Univ. New Mexico Pub. Geol. 5.

KENDALL, P. F., AND BRIGGS, HENRY, 1933, "The Formation of Rock Joints and the Cleat of Coal," Proc. Roy. Soc. Edinburgh, Vol. 53, pp. 164–87.

KIERSCH, GEORGE A., 1955, Mineral Resources (Navajo-Hopi Indian Reservations, Arizona-Utah), 3 Vols., Univ. Arizona Press, Tucson.

KINAHAN, G. H., 1875, Valleys and Their Relation to Fissures, Fractures and Faults, pp. 13, 17–23. Trubner & Company, London.

KING, PHILIP B., 1948, "Geology of Southern Guadalupe Mountains, Texas," U. S. Geol. Survey Prof. Paper 215, pp. 114–17.

KING, WILLIAM, 1880, "Preliminary Notice of a Memoir on Rock Jointing," Proc. Royal Irish Acad., 2d Ser., Vol. III (Sci.), No. 5, pp. 326–31.

LECONTE, JOHN, 1882, "Origin of Jointed Structure in Undisturbed Clay and Marl Deposits," Amer. Jour. Sci., 3d. Ser., Vol. 23, pp. 233–34.

LYELL, CHARLES, 1857, A Manual of Elementary Geology. Appleton and Company, New York.

McGEE, 1882, "Note on Jointed Structure," Amer. Jour. Sci., 3d. Ser., Vol. 25, pp. 152–53.

MATHER, JACKSON, AND HOUGHTON, 1841, "Comments on Jointng," Assoc. Amer. Geol., Acad. Nat. Sci., Philadelphia, Amer. Jour. Sci., 1st Ser., Vol. 41, pp. 172–73.

MELTON, FRANK A., 1929, "A Reconnaissance of the Joint Systems in the Ouachita Mountains and Central Plains of Oklahoma," Jour. Geol., Vol. 37, pp. 729–46.

NEVIN, C. M., 1949, Principles of Structural Geology, pp. 146–59. John Wiley and Sons, Inc., New York.

OROWAN, E., 1949, "Fracture and Strength of Solids," Reports on Progress in Physics, Vol. 12, p. 85.

PARKER, J. M., 1942, "Regional Systematic Jointing in Slightly Deformed Sedimentary Rocks," Bull. Geol. Soc. America, Vol. 53, pp. 381–408.

PHILLIPS, JOHN, 1855, Manual of Geology. Griffin and Company. London.

———, 1856, British Assoc. Report, p. 369.

RAGGATT, H. G., 1954, "Markings on Joint Surfaces in Anglesea Member of Demon's Bluff Formation, Anglesea, Victoria," Bull. Amer. Assoc. Petrol. Geol., Vol. 38, pp. 1808–10.

REPENNING, C. A., AND PAGE, H. G., 1956, "Late Cretaceous Stratigraphy of Black Mesa, Navajo and Hopi Indian Reservations, Arizona," ibid., Vol. 40, pp. 255–94.

ROBERTS, STEPHEN E., 1954, "Cracks in Asphalt Resurfacing Affected by Cracks in Rigid Bases," Nat. Research Council, Highway Research Board, Proc. 33d Ann. Meet., pp. 341–45.

SHOEMAKER, EUGENE M., 1953, "Collapse Origin of the Diatremes of the Navajo-Hopi Reservation" (abst.), Bull. Geol. Soc. America, Vol. 64, p. 1514.

STEWART, JOHN H., 1959, "Stratigraphic Relations of Hoskinnini Member (Triassic?) of Moenkopi Formation on Colorado Plateau," Bull. Amer. Assoc. Petrol. Geol., Vol. 43, pp. 1852–68.

SWANSON, C. O., 1927, "Notes on Stress, Strain, and Joints," Jour. Geol., Vol. 35, pp. 193–223.

TURNBOW, DIX R., 1955, "Permian and Pennsylvanian Rocks of the Four Corners Area," Four Corners Geol. Soc. Guidebook, pp. 66–69.

VAN HISE, C. R., 1895, "Principles of North American Pre-Cambrian Geology," U. S. Geol. Survey 16th Ann. Rept., pp. 668–72.

VER STEEG, KARL, 1942, "Jointing in the Coal Beds of Ohio," Econ. Geol., Vol. 37, pp. 503–09.

WAGER, L. R., 1931, "Jointing in the Great Scar Limestone of Craven and Its Relation to the Tectonics of the Area," Quar. Jour. Geol. Soc. London, Vol. 87, pp. 392–424.

WENGERD, S. A., AND STRICKLAND, J. W., 1954, "Pennsylvanian Stratigraphy of the Paradox Salt Basin, Four Corners Region, Colorado and Utah," Bull. Amer. Assoc. Petrol. Geol., Vol. 38, pp. 2157–99.

WOODWORTH, J. B., 1897, "On the Fracture System of Joints, with Remarks on Certain Great Fractures," Boston Soc. Nat. Hist., Vol. 27, pp. 163–83.

Copyright 1968 by The American Association
of Petroleum Geologists
BULLETIN OF THE AMERICAN ASSOCIATION OF PETROLEUM GEOLOGISTS
VOL. 52, NO. 1 (JANUARY, 1968), PP. 57–65, 5 FIGS., 1 TABLE

QUANTITATIVE FRACTURE STUDY—SANISH POOL, McKENZIE COUNTY, NORTH DAKOTA[1]

GEORGE H. MURRAY, JR.[2]
Billings, Montana 59102

ABSTRACT

The Devonian Sanish pool of the Antelope field has several unusual characteristics which make it almost unique in the Williston basin. Some of these are: (1) high productivity of several wells from a nebulous, ill-defined reservoir; (2) association with the steepest dip in the central part of the basin; (3) very high initial reservoir pressure; and (4) almost complete absence of water production.

Analysis of these factors indicates that Sanish productivity is a function of tension fracturing associated with the relatively sharp Antelope structure. Fracture porosity and fracture permeability can be related mathematically to bed thickness and structural curvature (the second derivative of structure). It is found that fracture porosity varies directly as the product of bed thickness times curvature and that fracture permeability varies as the third power of this product. A map of structural curvature in the Sanish pool shows good coincidence between areas of maximum curvature and areas of best productivity.

Volumetric considerations show that the quantities of oil being produced cannot be coming from the Sanish zone. It is concluded that the overlying, very petroliferous Bakken Shale is the immediate, as well as the ultimate, source of this oil. The role of the Sanish fracture system is primarily that of a gathering system for many increments of production from the Bakken.

The extremely high initial reservoir pressure indicates that the Sanish-Bakken accumulation is in an isolated, completely oil-saturated reservoir and, hence, is independent of structure in the normal sense. Similar accumulations should be present anywhere in the Williston basin where a permeable bed, of limited areal extent, is in direct contact with either of the two Bakken shale beds.

INTRODUCTION

The Sanish pool is one of several oil accumulations in the Antelope field of McKenzie County, North Dakota. As shown in Figure 1, this field is on a relatively sharp, southeast-trending anticline on the east side of the Nesson uplift of the central Williston basin. The field discovery, the Pan American No. 1 Woodrow Starr, SW¼ SE¼, Sec. 21, T. 152 N., R. 94 W., was completed in December 1953 with an initial flow potential of 550 bbl a day from 10,526–10,566 ft in the Devonian Sanish zone. Production in the Mississippian Madison Group was established in May 1956, and production in the Devonian Nisku and Duperow Formations and in the Silurian section was found in 1960. One well, the Amerada No. 1 Nelson, SW¼ SW¼, Sec. 5, T. 152 N., R. 94 W., recently has been recompleted as a discovery well in the Mississippian Lodgepole Formation. Cumulative production as of July 1, 1966 was 7,986,141 bbl from the Madison, 7,140,448 bbl from the Sanish, 1,072,890 bbl from the Duperow, and

1,477,410 bbl from the Silurian, a total of approximately 17 million bbl.

The Madison, Nisku, Duperow, and Silurian pools generally may be considered to be conventional structural accumulations. The Sanish pool, however, has several unusual characteristics which make it almost unique in the Williston basin. To date, the only other Sanish production in the United States part of the basin has been in the one-well, subcommercial Elkhorn Ranch field in Sec. 5, T. 143 N., R. 101 W., Billings County, North Dakota. Some of the very interesting aspects of the Antelope Sanish accumulation are a very high initial reservoir pressure, the high productivity of several wells from a nebulous, ill-defined reservoir, and, in contrast to most Williston basin fields, an almost complete absence of water production.

LITHOLOGIC AND RESERVOIR CHARACTERISTICS

SANISH ZONE

As shown in Figure 2, the so-called "Sanish zone" is at the very top of the Devonian Three Forks Formation, just below the lower Bakken shale. A thin, very dolomitic sandstone at the top of this interval was termed the "Sanish Sand" during the early development of the field. It originally was believed that the limits of Sanish production would be related to the areal extent

[1] Manuscript received, November 5, 1966; accepted, June 16, 1967. Modified from a paper presented before the 15th annual meeting, AAPG Rocky Mountain Section, Billings, Montana, September 28, 1965.

[2] Independent geologist.
The writer is indebted to W. H. Somerton, Univ. California, Berkeley, for checking the mathematical treatment presented herein.

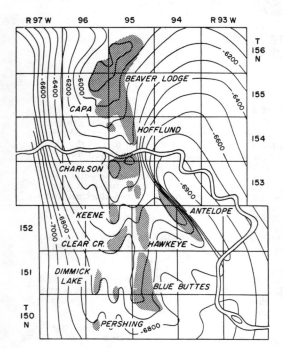

R 97 W   96   95   94   R 93 W

FIG. 1.—Regional structural contour map of McKenzie County area, North Dakota. Structural datum, base of lowest Charles (Mississippian) salt. Map shows location of oil pools on part of Nesson anticline. Contour interval, 100 ft. Contour datum, sea level.

and degree of development of this sandstone. However, subsequent drilling showed that this sandstone is absent in much of the field. Where the sandstone is absent, the Sanish zone consists of gray, very dense dolomite, commonly sandy, interbedded and intercalated with waxy, greenish, pyritic shale typical of the underlying Three Forks section.

Surprisingly, some of the wells with the best-developed Sanish sandstone have been among the poorest producers, whereas others with no sandstone have been among the best producers. For example, the No. 1 Reed-Norby Unit, NW¼ NE¼, Sec. 6, T. 152 N., R. 94 W., penetrated 4 ft of sandstone but has been a poor producing well. Cumulative production to July 1, 1966, was 73,876 bbl and productive capability at that time was less than 15 bbl a day. In contrast, the Carter No. 1 Norby-Melby Unit, SW¼ NE¼, Sec. 7, T. 152 N., R. 94 W., found no sandstone but has been one of the best wells in the field. Cumulative production to July 1, 1966 was 511,910 bbl and the

producing rate then was approximately 270 bbl a day.

Core analyses of the Sanish section indicate poor reservoir parameters regardless of the presence or absence of sandstone development. Porosity averages between 5 and 6 percent and plug permeability almost invariably is less than 0.1 md. Despite a universal absence of initial water production, core analyses indicate wide variations in oil and water saturations among wells.

BAKKEN FORMATION

The Mississippian Bakken Formation is composed of three easily differentiated units. The most striking of these are the two shale beds at the top and bottom of the formation, respectively. These are very radioactive, black, petroliferous shale and undoubtedly are source beds throughout the central Williston basin. They invariably have shows of oil and gas in cuttings and their log characteristics also are strong evidence for their identification as source beds. Their fluid-filled pore space, as indicated by velocity and/or neutron logs, is normal or greater than normal for shale at the depth at which they are found. Their

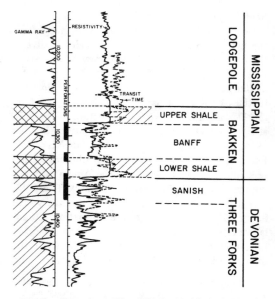

FIG. 2.—Logs from Lloyd H. Smith No. 1-A Weedeman, SE¼ SW¼, Sec. 29, T. 153 N., R. 94 W., McKenzie County, North Dakota, showing typical productive section. Curve on left is gamma ray. Solid curve on right is resistivity. Dashed curve is interval transit time. Depths (from surface) in feet.

resistivity, however, is essentially infinite. This fact, in conjunction with the visibly petroliferous character of the two shales, indicates that their pore space is hydrocarbon-saturated. The lower shale was perforated together with the Sanish section in several wells. However, no attempt has been made to separate the production by zones. Consequently, the direct contribution of the shale to the total production is unknown.

The section between the two shale zones is composed of dolomite, dolomitic siltstone, and minor quantities of shale. As might be expected, this section, called the "Banff member" by some workers, generally carries oil shows. This section also was perforated in conjunction with the Sanish zone in several wells in the field. However, as in the case of the lower shale, no attempt has been made to measure the production from the Banff independently of that from the Sanish.

LODGEPOLE FORMATION

The interbedded limestone and shale of the Mississippian lower Lodgepole Formation overlie the upper Bakken shale. This lower part of the Madison Group recently has been proved to be productive in a rework of the Amerada No. 1 Nelson, SW¼ SW¼, Sec. 5, T. 152 N., R. 94 W. Although not discussed here, this Lodgepole accumulation probably is related closely to the Bakken-Sanish pool.

ROLE OF FRACTURING IN SANISH PRODUCTION

THEORY

Because the productivity of individual Sanish wells appears to be unrelated to variations in reservoir lithology and because measured porosity and plug permeability are very low, it appears reasonable to assume that productivity is related to fracturing. Although core descriptions do not indicate an unusual degree of fracturing, a few fractures were noted. The data from the Carter No. 1 Norby-Melby Unit illustrate the peculiarities of the Sanish reservoir which lead to the hypothesis that fracturing is the controlling parameter. This well found no sandstone in the Sanish section—only dense dolomite and shale with some slight fracturing. Whole-core analysis shows an average porosity of approximately 5.5 percent and permeability below 1.0 md, except for a value of 27 md in a 1-ft zone and another value of 6.7

md in another 1-ft zone. Nevertheless, this well has been one of the best in the field! The writer believes that the only logical hypothesis to explain this fact is that the more than 500,000 bbl which has been produced from this well has come from fractures in the 2 ft of section with whole-core permeability values greater than 1.0 md.

The important role of fracturing also is suggested by the association of this production with the area of steepest dip in the central Williston basin. A cursory examination of the Sanish pool shows that the best wells are concentrated along the northwest-southeast line where the Antelope anticline bends abruptly from the relatively flat crest into the steep, northeast flank (Fig. 4). It is intuitively reasonable that the greatest intensity of tension fracturing would be in this position, where structural curvature (rate of change of dip, or structural second derivative) is greatest. Although this simple observation reinforces the hypothesis of a fractured reservoir, it indicates nothing about the magnitudes of fracture porosity and fracture permeability or about the minimum curvature necessary for the development of fractures.

To attempt to answer these questions, with particular reference to the Sanish pool and the individual wells in the pool, the writer has derived two mathematical relations which express fracture porosity and fracture permeability, respectively, as functions of bed thickness and structural curvature. Because the Antelope anticline is relatively elongate, the equations have been derived for a structural configuration which is infinite in the axial direction. This greatly simplifies the mathematical treatment.

Figure 3 shows a cross section of a segment of a competent bed, of thickness T feet, folded into an arc, of radius of curvature R feet. The folding is assumed to be sufficiently sharp to have caused stress greater than the ultimate tensile strength of the bed and, consequently, to have resulted in tension fractures represented by the idealized pie-shape voids. For convenience, the neutral surface (surface of no change in length, or no strain) has been taken as the base of the competent bed. The Z axis of Figure 3 is vertical; the Y axis—normal to the page—is chosen to coincide with the direction of the structural axis; and the X axis is the horizontal axis at right angles to the structure. The angle θ, measured in radians in a counter-

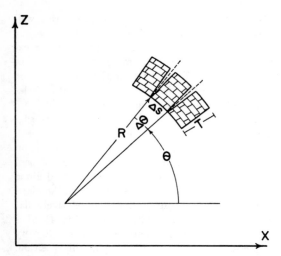

FIG. 3.—Geometry of fracture system used in deriving expressions for fracture porosity and fracture permeability.

clockwise direction from the positive X axis, is the angle made by a normal to the competent bed. The angular increment between adjacent fractures is represented by $\Delta\theta$. The corresponding increment in the surface of the fractured bed is represented by Δs.

The fractional porosity of the fracture system is determined easily from the geometry of the figure. Because the structural configuration and porosity are unchanging in the axial direction, the porosity may be calculated by considering a "slice" of the structure of unit length in the Y direction. Thus, from the expression for the area of a sector of a circle, the fractional porosity, ϕ, of that segment of the reservoir bounded by the two sides of angle $\Delta\theta$, the top and bottom of the bed, and unit length in the Y direction is given by

$$\phi = \frac{\frac{1}{2}(2RT + T^2)\Delta\theta - T\Delta s}{\frac{1}{2}(2RT + T^2)\Delta\theta}.$$

Because

$$\Delta s = R\Delta\theta,$$

this may be reduced to

$$\phi = \frac{T}{2R + T}.$$

T generally is very small in comparison with R. For example, the greatest thickness of any single competent bed, T, in the Sanish zone is on the

order of 10 ft, whereas the minimum value of R for the Antelope structure is approximately 10,000 ft. Hence, the last expression may be reduced to

$$\phi = T/2R. \tag{1}$$

It is useful to express the radius of curvature R in terms of the derivatives in the $X - Z$ coordinate system permitting the graphical evaluation of equation (1) from structural cross sections. In the usual structural situation, dz/dx, the dip measured in feet per feet is very small in comparison with unity. Hence, as is proved in most elementary calculus texts (Sherwood and Taylor, 1946), it is sufficiently accurate to take

$$R = 1 \left/ \frac{d^2z}{dx^2} \right. .$$

From this expression it may be observed that the curvature, which is defined as the reciprocal of the radius of curvature, is simply the second derivative of structure, d^2z/dx^2. Substitution into (1) yields the simple expression

$$\phi = \tfrac{1}{2}T\frac{d^2z}{dx^2} \tag{2}$$

for fractional porosity as a function of bed thickness and structural curvature.

It should be noted that fracture porosity generally is very small. For example, use of the maximum and minimum values of T and R, respectively, for the Sanish pool gives a maximum fracture porosity of approximately 1/2,000, or 1/20 of 1 percent.

As might be expected, fracture permeability is much more significant. The basic expression used in deriving the relationship between fracture permeability, bed thickness, and structural curvature is the equation for the volume of fluid flow per unit length between two parallel plates (Lamb, 1932),

$$q_p = - \frac{h^3}{12\mu}\frac{dp}{dy}$$

where h is the separation of the plates, μ is fluid viscosity, and dp/dy is the pressure gradient in the direction of flow. The minus sign arises because the direction of flow is in the direction of decreasing pressure. Because the angular divergence of the fracture faces is very small, the total volume of flow through one of the pie-shape fractures may be approximated by

$$Q = \int_0^T q_p dt = - \frac{1}{12\mu}\frac{dp}{dy}\int_0^T h^3 dt$$

where t is the distance in the fracture above the point of zero separation, or above the base of the fractured bed. For the pie-shape fracture, $h = kt$ and

$$Q = -\frac{k^3}{12\mu}\frac{dp}{dy}\int_0^T t^3 dt = -\frac{k^3 T^4}{48\mu}\frac{dp}{dy}.$$

The average flow per unit area is the total flow per fracture divided by the average area A per fracture,

$$q = \frac{Q}{A} = -\frac{k^3 T^4}{48\mu A}\frac{dp}{dy}.$$

In order to evaluate k, it is noted that

$$\frac{\frac{1}{2}h_0 T}{A} = \phi \quad \text{and thus} \quad k = \frac{A}{T^2}\left(T\frac{d^2z}{dx^2}\right),$$

where h_0 is fracture width at the top of the bed. Hence,

$$q = -\frac{A^2}{48 T^2 \mu}\left(T\frac{d^2z}{dx^2}\right)^3\frac{dp}{dy}.$$

From the last expression the permeability is seen to be given by

$$K = \frac{A^2}{48 T^2}\left(T\frac{d^2z}{dx^2}\right)^3.$$

After conversion from cgs units to the more familiar millidarcy (Pirson, 1950) and evaluation of A in terms of an assumed fracture spacing of 6 in., this expression reduces to

$$K = 4.9 \times 10^{11}\left(T\frac{d^2z}{dx^2}\right)^3. \tag{3}$$

Although the assumption of a fracture spacing of 6 in. admittedly is arbitrary, there are certain considerations leading to this as a reasonable approximation. If the fracture spacing were much greater than this it would be possible for some of the favorably located wells to have missed the fracturing altogether. There is no evidence of this in the Sanish pool. Also, assumption of wider spacing leads to unreasonably large values of calculated permeability. Finally, the writer's field experience leaves the impression that a 6-in. fracture spacing may be somewhat short, but that it is not entirely unreasonable.

Because permeability increases as the third power of $T\ d^2z/dx^2$, it becomes appreciable with relatively small values of this parameter. The tabulations in Table I give illustrative permeability values for different values of T and d^2z/dx^2.

TABLE I. EXAMPLES OF PERMEABILITY VALUES

$T = 5$ ft		$T = 10$ ft	
$\frac{d^2z}{dx^2}$ ($\times 10^{-5}$)	K (md)	$\frac{d^2z}{dx^2}$ ($\times 10^{-5}$)	K
1	0.06	1	0.49
2	0.49	2	3.92
4	3.92	4	31.20
6	13.20	6	106.00

The figures in Table I should not be taken too literally. The configuration of a natural fracture system undoubtedly varies considerably from the assumptions made in deriving the above expressions for porosity and permeability. Particularly subject to question are the assumptions that the base of the competent bed is a neutral surface and that the fracture spacing is 6-in. It also is apparent that the regular, pie-shape fracture voids are an oversimplification. In some cases the value of bed thickness is uncertain. T represents the thickness of a single competent bed unbroken by shale intercalations capable of providing slippage. In a section like the Sanish, with abundant shale intercalations, a value for T is difficult to ascertain. Despite these uncertainties, it is believed that the above expressions are important as indications of the order of magnitude of fracture porosity and fracture permeability which might be expected with a particular bed thickness and structural configuration.

It should be noted that a minimum value of the parameter $T\ d^2z/dx^2$ must be exceeded before fracturing is developed. It is easily shown that the tensile stress in the upper surface of the bed of Figure 3 is given by

$$F = ET\frac{d^2z}{dx^2}$$

where F is the stress in the upper surface and E is Young's modulus for the bed (Stephenson, 1952). If $T\ d^2z/dx^2$ is such that F exceeds the ultimate tensile strength of the bed, fractures will develop. A rough idea of this critical value may be obtained from measured values of the ultimate tensile strength and Young's modulus of building stones. Some average values (Kidder and Parker, 1935) would give a critical value of about 1.2×10^{-4} for $T\ d^2z/dx^2$ in the case of a limestone. As discussed hereafter, the empirically determined critical value of structural curvature, d^2z/dx^2, for the presence of Sanish fracturing appears to be about 2×10^{-5}. Depending on the exact value of T, this would place $T\ d^2z/dx^2$ near the above figure of 1.2×10^{-4}.

APPLICATION TO SANISH POOL

Figure 4 shows the correlation between well productivity and structural curvature in the Sanish pool. Figure 5 shows how the values of curvature on the map were obtained. At the top of Figure 5 is a structural profile of the Bakken Formation along section A-B. The central profile is a plot of dip magnitude along this line of section. This second curve has been obtained from the first by drawing tangents to the first curve, as indicated, measuring the dip in feet per hundred feet, and plotting this magnitude at the position measured. West dip has been chosen arbitrarily as positive and east dip as negative. The bottom profile of structural curvature has been obtained from the second curve just as the second was obtained from the first. Tangents are drawn on the dip-magnitude curve, their slopes measured, and the values of these slopes plotted at their respective positions.

Across the Antelope field from west to east, the curvature increases to a maximum of about 5×10^{-5}/ft where the dip is changing abruptly into the steep east flank, then returns to zero at the inflection point in the middle of the east flank, and reaches a second maximum where the beds bend sharply into the syncline on the east. As indicated in the legend in Figure 4, the dotted areas show where the curvature is between 2×10^{-5}/ft and 4×10^{-5}/ft and the dashed areas show where the curvature is greater than 4×10^{-5}/ft. As outlined above, the minimum curvature necessary for the development of tension fractures in a 10-ft bed is approximately 2×10^{-5}/ft. Hence, as a first approximation, production should be restricted to the dotted and dashed areas. This appears to be empirically correct. The few producing wells in areas where the structural curvature is less than 2×10^{-5}/ft are all subcommercial.

The matter of sign is arbitrary. Downward curvature has been taken as negative and upward curvature as positive. However, the direction and sign of the curvature in a few cases could be significant with regard to fracture geometry. The widest part of a fracture is at the top of the competent bed if the curvature is downward. If the curvature is upward, the widest parts of the fractures are at the base of the bed.

After several similar cross sections were constructed, the areas where the curvature exceeds the critical value for the development of fracture permeability (dotted) and the areas where the curvature and fracture permeability are the greatest (dashed) were mapped. As indicated in the legend in Figure 4, the well spots are keyed to their productivity. The match between actual productivity and the theoretical fairway is not perfect, as shown by the presence of some of the best wells and a few very poor wells in the dotted areas. However, if the statistical variation which is inevitable in a fracture reservoir is considered, the match between theory and experience is remarkably good.

SANISH PRODUCTION AND RESERVES

There is a wide variation in productivity of the Sanish wells. Prior to recent drilling on the north end of the field, there were 14 wells along the fairway of maximum structural curvature which had produced most of the oil from the pool. Each of these 14 wells, as of January 1, 1966, had produced more than 250,000 bbl and together had produced 74 percent of the oil recovered from the pool. The average producing rate of each well was 178 bbl a day at that time. The other 19 wells had produced the other 26 percent of the total. The 16 of these 19 poorer wells which were still producing as of the first of 1966 had an average producing rate at that time of only 17 bbl a day apiece.

The five northernmost wells in the fairway have been drilled during the past $1\frac{1}{2}$ yr. Although there is some variation in the productivity of these recently completed wells, their average is as good as that of the best 14 older wells.

It is impossible to derive a meaningful reserve estimate for the Sanish from volumetric calculations. The better wells already have produced several times as much oil as could be estimated for the ultimate recovery from core analysis. For example, the discovery well, the No. 1 Woodrow Starr, SW¼ SE¼, Sec. 21, T. 152 N., R. 94 W., had a January 1, 1966, cumulative production of 607,467 bbl. (This includes production from the Starr No. 1-A which was drilled on the same 160-acre location after the casing collapsed in the original well.) The 14 best wells had a cumulative

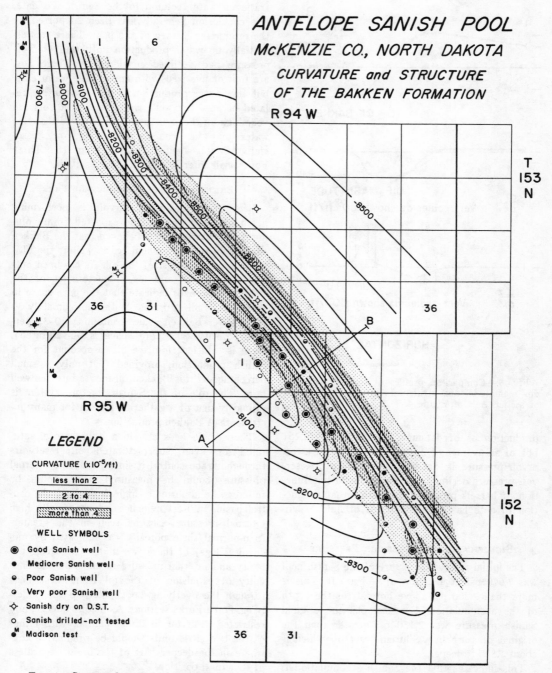

FIG. 4.—Structural contour map of Antelope Sanish pool, McKenzie County, North Dakota. Structural datum, top of Mississippian Bakken Formation. As noted on legend, well spots are keyed to their productivity and values of structural curvature are mapped by patterned areas. Contour interval, 50 ft. Contour datum, sea level. Section A-B is location of Figure 5.

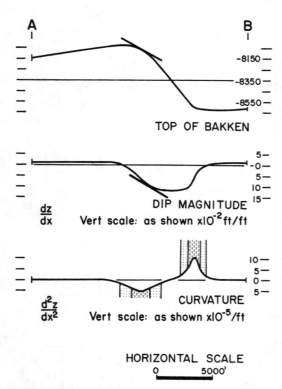

TOP OF BAKKEN

$\dfrac{dz}{dx}$ DIP MAGNITUDE

Vert scale: as shown $\times 10^{-2}$ ft/ft

$\dfrac{d^2z}{dx^2}$ CURVATURE

Vert scale: as shown $\times 10^{-5}$/ft

HORIZONTAL SCALE

0   5000'

FIG. 5.—Comparison profiles of Bakken structure, dip magnitude, and structural curvature along line of section A-B of Figure 4.

production as of January 1, 1966 of 4,940,261 bbl, or an average of 2,200 bbl/acre. Average reservoir pressure at that time was 3,644 psi. Reservoir-pressure decline has been a straight-line function of barrels produced, indicating that production to date has been the result of fluid expansion.

SIGNIFICANCE OF RESERVOIR PRESSURE

The initial reservoir pressure of the Sanish pool was 7,670 psi at a datum of −8,400 ft. This is more than 2,500 psi above normal for the depth of the accumulation. For example, the original Sanish pressure was 2,150 psi greater than the original pressure in a Silurian reservoir which is about 1,500 ft deeper.

This indicates that the Sanish accumulation is in a closed, completely oil-saturated reservoir. The most important consequence of this fact is that the Sanish pool is independent of structure in the normal sense and there is no risk of pene-

trating a water column in the Sanish. To date, this conclusion generally has been supported by the productive history of the pool. There was essentially no water production until the reservoir pressure had declined considerably. There now are four or five wells with appreciable water cuts, but this water production is not structurally related because the wells with the largest cuts are very high on the structure. It is believed that this water may be true connate water—or "associated" water—being expelled from the Three Forks shale interbedded with the reservoir.

BAKKEN SHALE AS SOURCE OF OIL

Because the Sanish reservoir is performing much better than could be expected from volumetric calculations, it is believed that the Bakken shale beds are the immediate, as well as the ultimate, source of most of the oil. As has been outlined, the visual and log characteristics of the Bakken shale beds indicate that the pore space in the shales is oil-saturated. With approximately 30 ft of lower Bakken shale, containing 30–35 percent porosity (the pore space is oil-saturated), there is sufficient in-place oil to account for the Sanish production, provided that only a small percentage of this Bakken oil moves into the well bore. The role of the Sanish fracture system is primarily that of a gathering system for many increments of Bakken production.

These shale beds are, in a larger sense, saturated or even supersaturated oil reservoirs throughout the central basin area. If the internal pressures could be measured, they would be found to be abnormally high throughout the central basin. Indirect evidence for abnormally high internal pressures can be seen in the greater-than-normal shale porosity (as indicated by mechanical logs) of these two units. The excess porosity and internal pressure result from the difficulty of expulsion of the oil from these beds through the overlying Lodgepole and the underlying Three Forks sections. As a consequence, any restricted reservoir in direct contact with either of the two shale units should be productive anywhere in the deeper part of the basin, regardless of structural position.

CONCLUSIONS

Analysis of the Sanish pool supports the immediate impression that the production is from a tension-fracture system associated with the relatively sharp Antelope fold. In the local context of the Williston basin, one of the most important conclusions is the recognition that the upper and lower Bakken shale beds are supercharged oil shales and that they probably are the immediate source of most of the oil. Hence, discovery of permeability in beds adjacent to either of these two shale units—either in another fracture system or an area of matrix permeability in sandstone or carbonate rocks—would probably mean the discovery of prolific production comparable with that of the Sanish pool. Equally intriguing is the conclusion that, because the two shale beds are everywhere oil-saturated, such production is independent of structure in the normal sense and might be found in a synclinal location. Most important, it is believed that the relations between fracture porosity, fracture permeability, bed thickness, and structural curvature may prove useful in the analysis of similar problems in other geologic provinces.

REFERENCES CITED

Kidder, F. E., and compiled by a staff of specialists and Harry Parker, editor-in-chief, 1935, Kidder-Parker architects' and builders' handbook: 18th ed., New York, John Wiley and Sons, Inc., p. 296.

Lamb, Sir Horace, 1932, Hydrodynamics: 6th ed., New York, Dover Publications, p. 581–582.

Pirson, S. J., 1950, Elements of oil reservoir engineering: New York, McGraw-Hill Book Company, Inc., p. 47.

Sherwood, G. E. F., and Angus E. Taylor, 1946, Calculus: rev. ed., New York, Prentice-Hall, Inc., p. 206–209.

Stephenson, R. J., 1952, Mechanics and properties of matter: New York, John Wiley and Sons, Inc., p. 245–246.

Copyright 1968 by The American Association
of Petroleum Geologists
THE AMERICAN ASSOCIATION OF PETROLEUM GEOLOGISTS BULLETIN
VOL. 52, NO. 3 (MARCH, 1968), P. 420-444, 38 FIGS.

EXPERIMENTAL ANALYSIS OF GULF COAST FRACTURE PATTERNS[1]

ERNST CLOOS[2]
Baltimore, Maryland 21218

ABSTRACT

Gulf Coast faults are normal faults with the exception of those around salt domes. Normal faults must accommodate both a horizontal and vertical component of displacement. The horizontal component increases the distance between two points on opposite sides of the fault surface. The vertical component is larger and must be accommodated simultaneously by the same mechanism. Experiments show that horizontal and vertical components can be derived from one motion in practically horizontal surfaces.

The accumulation of all horizontal components in the Gulf Coast embayment makes considerable horizontal displacement of the sedimentary blanket necessary.

The regional Gulf Coast fault pattern and its many local variations are therefore thought to be caused by regional gravity creep of the sedimentary blanket into the basin. As creep takes place the sliding sediments break away from the stationary ones forming a marginal graben, and, nearer the coast, asymmetrical down-to-basin faults with reverse drag, and antithetic faults.

Creep has taken place for a long time as is proved by the fact that there was sedimentation after and during faulting in association with many growth faults. The first faults may well have been the peripheral ones as they occurred first in the experiments. Faults within the basin are later, as shown by the facts that they were buried under sediments, grew upward into them, and transect younger formations.

Local domes show local fault patterns, but at many places the regional fault pattern suppressed local ones. Experimentation suggests that a general mechanism can explain both regional and local phenomena rather simply by one general cause which is modified by specific local conditions within the large and heterogeneous Gulf Coast embayment.

INTRODUCTION

Faults and fault patterns in the coastal plain bordering the Gulf of Mexico have been described comprehensively by Murray (1961, chaps. 4 and 5). With very few exceptions "normal (gravity-tensional) faults predominate" (Murray, p. 167; see Fig. 1, this paper).

Some of these normal faults are caused by regional stresses which result in regional fault patterns, and others are due to local stresses such as those generated by the rise of salt domes. Interference between these two patterns is common.

The faults can be classified also according to

[1] Manuscript received, April 10, 1967; accepted, July 31, 1967.

[2] Professor of Geology, The Johns Hopkins University, and consultant, Esso Production Research Company.

The late H. N. Fisk, former Professor of Geology at Louisiana State University and later Chief of the Geological Research Section, Humble Oil and Refining Company, now Esso Production Research Company, Houston, suggested an experimental study of Gulf Coast faulting and especially down-to-basin faults. The work was done at the Geological Laboratory, The Johns Hopkins University. I am grateful to the Esso Production Research Company for permission to publish this paper and for much stimulating discussion at the Esso laboratory.

I am grateful also for much needed editorial help by readers and critics. Illustrations were prepared by Josephine Spemann, Ulf Wiedemann, and Charles Weber.

special characteristics:

1. Normal faults which limit a graben such as are prominent in the inner zone of the Gulf Coast embayment: Balcones-Luling, Mexia-Talco, South Arkansas, and Pickens-Gilbertown fault systems;

2. Down-to-basin faults with reverse drag (sags, rollovers) are most common near the coast;

3. Antithetic faults;

4. "Growth faults" (Ocamb, 1961) or "contemporaneous faults" (Hardin and Hardin, 1961) with increased sedimentary thickness on the downthrown side; and

5. Faults related to salt domes.

There are gradations and overlapping characteristics among the different types of faults as well as combinations, which make it difficult and artificial to separate them. Therefore I have tried to design models for typical situations such as a graben, a down-to-basin fault with reverse drag and antithetic faults, and a salt dome with radial faults. After typical situations were reproduced I tried to imitate the regional pattern of Gulf Coast faulting in one model, with moderate success.

EXPERIMENTAL METHOD, PURPOSE, AND PROCEDURE

The experimental method used here has been described by H. Cloos (1928, 1930, 1931) and E. Cloos (1955) and is based on the fact that soft clay fractures when subjected to stress.

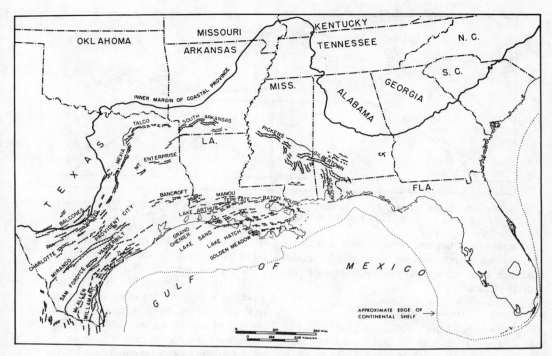

FIG. 1.—Diagrammatic representation of principal strike-fault systems in northern Gulf Coast (from Murray, 1961, Fig. 4.1; published with permission of Harper & Row, New York).

It is easy to imitate fracture patterns in clay models. The value of such imitation lies not in quantitative determinations of parameters but in the imitation of a movement pattern or stress pattern which is thought to be the cause of fractures or fracture patterns, or in the correction of erroneous notions which do not explain an observed pattern. In the following pages experiments are described for each geological situation which was studied. Each experiment was repeated many times and improved as errors were eliminated. For details on the experimental method, see E. Cloos (1955).

GENERAL CONSIDERATIONS

Because Gulf Coast faults are normal faults their displacements have a vertical and a horizontal component. Figure 2A shows this on a cross section and Figure 2B shows a map view of the horizontal component and the fault trace. The horizontal component means an extension between two points on opposite sides of the fault plane. On curved faults the horizontal component is less at upper levels and larger at lower and

gently dipping levels. This fact is important in the interpretation of curved normal faults and reverse drag as shown subsequently.

In most cross sections of Gulf Coast faults the horizontal component is underemphasized in comparison with the vertical component by reason of the gross vertical exaggerations which distort the geological profile (see Murray, 1961, Figs. 4.22b, 4.25, 4.27, 4.31, and others). Exaggerations are

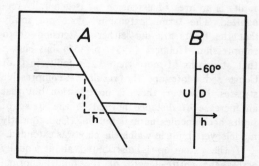

FIG. 2.—A, Profile of normal or gravity fault with horizontal components h and vertical component v. B, Map view of 2A.

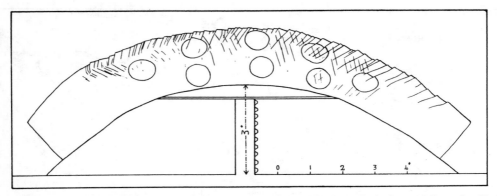

Fig. 3.—In bent beam, extension is distributed through length of beam with faults spaced evenly. Vertical uplift of clay model is carried by stiff but flexible metal sheet.

commonly used because sections with a scale of 1:1 do not permit showing stratigraphic details, although such sections do convey a far more accurate picture.

Graben Structures

As shown in Figure 1 (Murray, 1961, Fig. 4.1) the area which contains the "principal strike fault systems in northern Gulf coastal province" is bounded on the west, north, and east by the Balcones-Luling, Mexia-Talco, South Arkansas, Pickens, and Gilbertown graben systems. The Mexia-Talco graben system consists of smaller *en échelon* grabens but the remaining systems are almost uninterrupted. The Gulf Coast graben system was discussed recently by Walthall and Walper (1967), who compared it with African and European systems (p. 102) and concluded that these grabens are the result of post-orogenic movements of the Ouachita belt.

A graben bounded by two or more normal faults is an area of extension, as is generally recognized. The term "extension" does not imply that the faults are due either to tension or to compression. Hubbert (1951, p. 367) has shown that at depths of more than 30 ft faults in Gulf Coast sediments are the result of compressive stresses. However, there is no question but that an increase of distance normal to the strike of faults accompanies normal faulting. Consequently models were built in which the clay was extended.

If the extension is distributed about equally across the entire length of the model, normal faults will form that strike normal to the direction of extension and dip from 40° to 70° (*see*

also Hubbert, 1951). In profile the dip of the faults will be either to the left or right.

Such a setup was described by H. Cloos (1930, p. 741) and E. Cloos (1955, p. 246) and can be produced readily by modeling the clay on a one-way stretch belt, or on a rubber bladder, or by bending a beam by doming as shown in Figure 3.

A graben is a special case of such extension and can be produced experimentally by placing a clay block over two metal sheets or pieces of plywood which then are pulled apart. The clay block is not extended across its entire length but only above the joint between the base plates; then shear fractures appear at the separation point and grow upward. A graben forms between the fractures as the center block sags downward.

This experiment has been repeated many times and most recently it was pictured by Badgley (1965, p. 168, Fig. 5-17). H. Cloos (1953, Pl. XLIV) pictures the same experiment in a drawing. The extension here was produced by doming (see also H. Cloos, 1949, Fig. 56, facing p. 225). A similar graben caused by lateral pull of two separating base plates is shown by H. Cloos (1936, p. 265). Finally the same experiments are illustrated by Hills (1963, p. 187, Figs. VII-29, A and B).

Reverse Drag

To imitate reverse drag on down-to-basin faults, I began experimentation with a setup which provided maximum symmetry: a clay model, pulled equally in opposite directions, with a gap opening between the base plates of brick or wood blocks as shown in Figure 4. This experi-

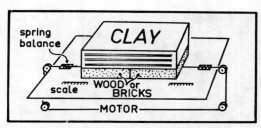

FIG. 4.—Diagrammatic setup of experiment for symmetrical pull and movement.

ment also has been described by H. Cloos (1930, p. 744) but not for the same reason. In that experiment he tried only to produce a graben; my experiment was aimed at maximum symmetry in a graben.

Symmetrical pull was maintained by spring balances on which the pull could be measured within 0.5 lb or less. The rate of movement was observed on a scale which showed the displacements of the moving base plates. Even with equalized pull, movements could become unequal due to differences in friction between the base and the table top, if, for example, one side "stuck" due to spilled clay.

Figure 5 shows the front of a model on blocks before movement began. "Bedding" is scored on the front surface of the model; the clay is homogeneous.

Figure 6 shows the formation of a symmetrical graben. Bedding remains horizontal in the center and is dragged slightly along the faults. Displace-

ment is equal at the two faults. On the left side one fault is prominent; on the right side antithetic faults appear in a zone.

The horizontal displacements of the blocks are equal within 3 mm of the 20-cm mark in the center. The pull on two scales was 16 lb on each side when fracturing began. Fractures first formed at the bottom where the blocks separated. Before they reached the surface plastic deformation occurred in the upper half of the model, and fracturing in the lower half. The curvature of the faults (in cross section) was caused by plastic sag in the upper part of the model before fracturing, thereby steepening the position of the shear planes.

Clay-particle movements were traced from successive photographs, as shown in Figure 7. Particles were identified and traced on acetate sheets and their paths are shown as dashed lines. In the footwalls particles moved horizontally away from the center at equal rates. Near the top of the model, particles first sagged due to plastic deformation and then moved horizontally away from the center. Within the graben the particles moved downward; the subsiding block spread and the horizontal distance between reference points increased slightly. The faults also moved laterally and thus opened the gap into which the block dropped. The movements were symmetrical within the limits of the model: away from the center in the footwalls and downward with a slight tendency toward the left.

FIG. 5.—Clay model on wooden blocks before movement. Scale in cm.

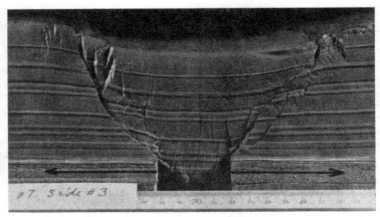

FIG. 6.—Symmetrical graben after symmetrical pull. Clay graben has slid down between blocks. Displacement is 20 mm on right side and 22 mm on left side. Scale in cm.

For a symmetrical graben caused by one-sided pull, a second model was designed identical with the first except that only the right block was pulled while the left block remained stationary. This experiment showed identical fracture patterns except for small details, regardless of whether the pull was toward left or right.

Figure 8 appears symmetrical and seems identical with the first experiment (Fig. 6). A difference is detected only when particle movements are traced (Fig. 9).

The left boundary is an inclined plane. The subsiding graben block slides along that plane, and the entire block can move only parallel with that boundary surface. It can do this only if the right block makes room for it by moving horizontally toward the right. There are no suggestions of downbending. Thus, it is evident that the movement alone is not the cause of asymmetries.

A symmetrical graben can be modeled also without the gap into which the block drops. The experimental setup is the same except that the base of the graben block impinges on the table top and this results in a zone of complex deformation. In such a model the graben block commonly breaks up into a group of subsidiary grabens and horsts, and the structure does not resemble the down-to-basin faults of the Gulf Coast (see H. Cloos, 1930; 1936, p. 265; 1949, Fig. 56; 1953, Pl. XLIV; Badgley, 1965, p. 168; Hills, 1963, p. 187).

DOWN-TO-BASIN FAULTS

All down-to-basin faults in the Gulf Coast are essentially asymmetrical graben structures in which the gulfward side has become a downbend (reverse drag, rollover) with synthetic and antithetic faults, and the updip side is bounded by a master fault that dips toward the basin. At many places stratigraphic thicknesses on the downthrown side are greater than on the upthrown side. Such faults have been called "growth faults" (Ocamb, 1961) or "contemporaneous faults" (Hardin and Hardin, 1961).

The dip of the master fault seems to lessen with depth (Murray, 1961, p. 182). This has been observed directly in the field by Hamblin (1965) who discussed most ably the phenomena of reverse drag and asymmetric faults.

A good example of a down-to-basin fault with reverse drag from the Isthmus of Tehuantepec is

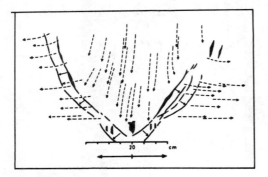

FIG. 7.—Migration of reference points during deformation. Traced from successive photographs.

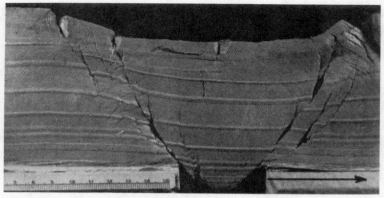

FIG. 8.—Symmetrical graben formed by asymmetrical pull toward right. Final stage: bedding remains horizontal, drag on faults is normal; dip slip on faults is equal. Scale in cm.

shown in Figure 10 (Castellot and Caletti, 1958). Here the Gulf of Mexico is on the north and the structure is a mirror image of Texas and Louisiana down-to-basin faults. In Figure 10 there is no well control (vertical lines) for the lower two horizons, but the upper four controlled layers show the reverse drag very well.

It has been shown already that the direction of tectonic transport is not the cause of the asymmetry of down-to-basin faults. To find the reason for the asymmetry a model was built on two overlapping metal sheets which move across the table top (Fig. 11).

Five stages of the experiment are shown in Figures 12–16. The right sheet overlaps the left one, as shown by arrows; the shaft of the upper arrow ends at the edge of the upper sheet. The left arrow indicates only the direction of motion of

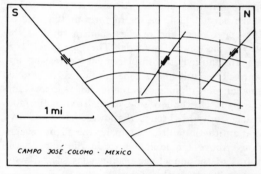

FIG. 10.—Reverse drag in generalized cross section of José Colomo field, Isthmus of Tehuantepec, State of Tabasco, Mexico (after Castellot and Caletti, 1958, Fig. 2). Vertical lines are wells. Lower two layers are uncontrolled, but controlled layers show structure.

the lower sheet. The rate of motion is equal for the two sheets within the limits of the experiment.

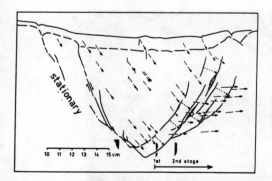

FIG. 9.—Migration of reference points during unequally pulled experiment. If pull is toward left, picture would become mirror image.

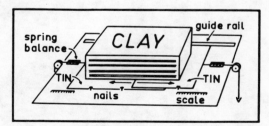

FIG. 11.—Experimental setup for asymmetrical fault pattern with symmetrical pull. Metal sheets overlap. Nails and metal rail prevent rotation of metal sheets. Spring balances permit equalizing pull. Movements are measured as displacements of metal sheets on table top.

Fig. 12.—Clay model on overlapping metal sheets before deformation. Left end of upper arrow is at edge of right, upper sheet.

After only a few millimeters of lateral displacement, fractures appear. At stage 2 (Fig. 13) a sharp normal fault begins to form at the edge of the upper sheet and reaches about a third of the way to the top of the model. The four lowest layers are faulted, the displacement decreasing upward. The fault also steepens upward. A second, smaller, left-dipping, synthetic fault appears on the right. On the left side, a small synthetic fault reaches upward through the lowest layer and several antithetic faults accompany the slight reverse drag. Above the faulted area in the upper two-thirds of the model the bedding has sagged symmetrically above the graben.

In Figure 14 faults reach the surface, but the upper two layers are not displaced. Fractures have just begun to form near the top. Near the base displacements on the right have grown larger, and the small fault on the right extends higher although not to the surface. Antithetic

faults at the left have become more extensive; the fault zone has become wider. Downbending is asymmetrical toward the right in the lower three layers.

At this stage a master fault at the right, a fault zone at the left, and reverse drag in the lower half of the model are well established. The master fault begins at the edge of the upper sheet and steepens toward the top of the model.

In Figure 15 the two sides of the graben are distinctly different. Downbending, especially in the lower half of the model, is very pronounced and is accomplished in part by plastic deformation and in part by rotation of blocks on antithetic (as related to the downbend) left-dipping faults in a zone that grows wider as deformation continues. A few synthetic right-dipping faults also appear near the top in the left fault zone; horst and graben patterns occur in the upper half of the model.

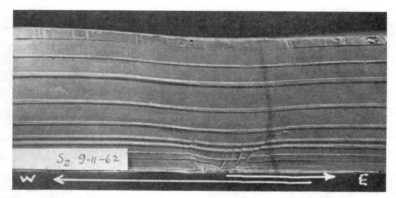

Fig. 13.—Asymmetrical graben after deformation begins. Faults appear first at base and downbending develops in lowest layers.

FIG. 14.—Faults at right and fault zone at left have reached surface. Asymmetry grows more pronounced. Downbending grows upward and includes upper layers.

Figure 16 is the final stage of the experiment. The block on the right has moved as a unit. The fault at the left has widened to include the entire area of downbending. Antithetic faults prevail; blocks have rotated clockwise, intensifying the downbending. In the upper part of the model faults dip left and right accommodating the flexure. Next to the master fault, bedding in the graben is still almost horizontal.

The movements of particles (Fig. 17) indicate the horizontal outward movement of the two blocks toward the left and right away from the central area. Within the graben the particle paths are asymmetrical. They move downward only nearest to the master fault. In the left half of the graben particles first move down and then toward the left as, for example, points A and F.

Line A-B-C dipped 4° at the beginning and line A'-B'-C' dipped 24° at the end of the experiment. During the same time interval the line extended from 26 mm to 38 mm. This illustrates the development of a downbend toward the master fault and of an extension which can be accommodated only by plastic flow and additional faulting. As the downbend develops the antithetic faults flatten. Line F-G shows the same relationship; an almost horizontal line becomes tilted and longer as the points migrate from F and G to F' and G'. Figure

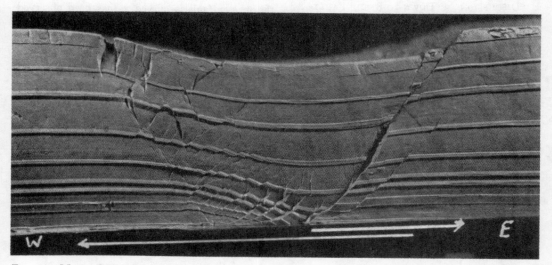

FIG. 15.—Master fault at right separates right block from asymmetrical graben. Left fault zone is wide and accommodates downbending in lower layers, where blocks between antithetic faults rotate clockwise.

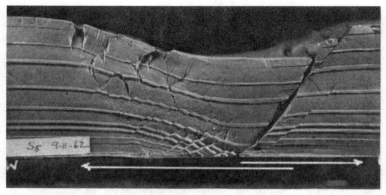

FIG. 16.—Antithetic faults dip left and blocks rotate in downbend.
Master fault is sharp and prominent, flattening downward.

18 pictures an essentially similar experiment in three dimensions. The right block was moved toward the right. A master fault is covered with striae; the downbend shows mostly antithetic faults.

Asymmetry is not a function of the motion of the two blocks, but the underlapping sheet removes the lower layers of the subsiding graben as they slide down the master fault plane onto the sheet. An identical asymmetry, with similar faulting and downbending, is obtained if pull is in one direction only. In such cases, particles move only away from the master fault which remains stationary.

DISCUSSION OF DOWN-TO-BASIN FAULT EXPERIMENT

The experiment essentially imitates what is

seen in field observations of down-to-basin faulting: a master fault, reverse drag, and antithetic faults. In addition, there is some thickening on the downthrown side, and the dip of the beds above the glide plane steepens. The master fault dips more gently at depth and, due to rotation, there seems to be less displacement above than below.

If one visualizes the right side (master fault) of Figure 16 as stationary and the left side as moving away from it, the structural pattern is produced by gliding on a surface alone without need for salt ridges (Quarles, 1953), faulted anticlines, basement participation, differential compaction, or any other mechanism. Furthermore, the master fault is curved as has been observed by Hamblin (1965). The experiment strongly supports the slipping-plane hypothesis as outlined by

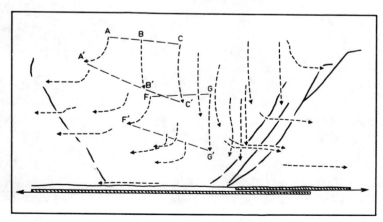

FIG. 17.—Migration of reference points in asymmetrical graben experiment (Figs. 11–16). Line A-C rotates into position A'-C', line F-G becomes F'-G'. Traced from successive photographs.

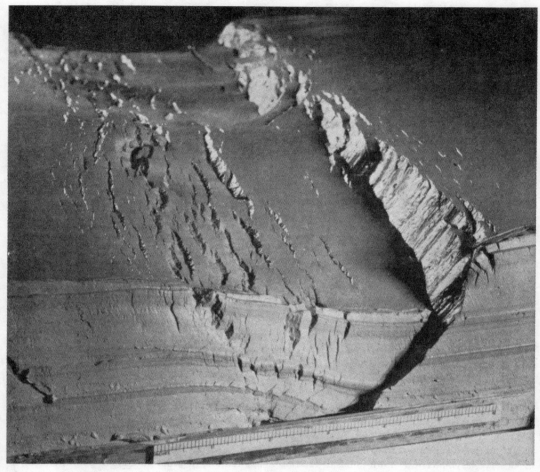

FIG. 18.—Reverse drag, master fault with striae, antithetic faults in oblique photograph. Right block moved right; left block remained stationary.

Quarles (1953, p. 498–500).

A creep mechanism can produce all the elements which have been observed on down-to-basin faults with reverse drag but only if the material at the base of the graben creeps first down the master fault and then along the glide plane. When faults first form in the symmetrical graben, both the updip and the down-to-basin shear planes exist, but the latter will develop into a master fault and the updip plane degenerates into a downbend with tension and a series of either synthetic or antithetic faults, or both. Where the material does not creep downdip and no rotation occurs, the updip fault, as well as the downdip fault, can become the master fault. This happens in the peripheral graben and is discussed in a subsequent section. It also happens in a few places

near the coast, and occurred in the experiments described below.

The experiments suggest that reverse drag should be more common and more intense on down-to-basin faults and rarer or much less intense on updip faults.

DOWN-TO-BASIN FAULTS AND SEDIMENTATION (GROWTH FAULTS)

In experimentation with clay, fractures are observed on free surfaces of the model. These surfaces are most sensitive and easily destroyed. The experimental material therefore was changed from clay to sand. The addition of sand to a sand model does not disturb existing structures. Also, glass walls provide a clean and even cross section. Sand moves along glass with little friction, and

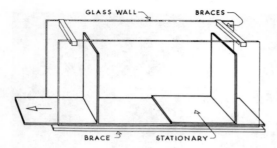

FIG. 19.—Setup for growth fault experiment. Partitions and floors are ¼-in. plywood. Glass plates are ¼-in. plate glass. Length of box 52 in., width 7 in. (schematic, not to scale).

the glass influences the fault pattern very little.

A long box was constructed of two plate glass walls, with a stationary end, and a movable end (Fig. 19). The movable end (left) was rigidly connected to a movable floor which reached almost the length of the box. The stationary end (right) was connected to a stationary floor above the movable one and extended to the center of the box. Sand used ranged from very fine to 2 mm in diameter.

Sedimentation was simulated by filling sand into the graben as it formed so that the slope of the sand surface did not intersect the fault which continued to grow upward as deformation continued. Sand was replenished as soon as the fault became visible as a feeble line at the surface.

The sand experiment served two purposes: (1) to permit sedimentation during faulting, and (2) to observe asymmetrical down-to-basin faulting. However, for the latter purpose, clay is very much better.

In the first stage of the experiment (Fig. 20) faulting has affected all but the topmost layer.

The second layer from the top, below the dark line, has been faulted, and more sand was added to it during faulting; the layer is thicker in the graben than on the two sides. The top layer also is faulted and was filled in after considerable faulting. The right (left-dipping) fault is steeper than the left one. Most of the layers were lowered vertically and remained horizontal between the two major faults. The lowest white layer and the darker one below it show pronounced asymmetry and downbending.

In the second stage (Fig. 21) all layers are faulted and the right fault is sharp and does not change dip. The left side is more asymmetrical and several faults appear in the left half of the graben. The dip of the left fault is gentler than that of the right fault.

The positions of both major faults (as they were in stage 2; Fig. 21) were marked by black lines on the glass (Fig. 22) to show changes of dip or position which resulted from further movement. The right fault remained stationary. The left fault moved toward the left with the entire moving block and several smaller faults appeared within the graben. As movement continued, there was downbending toward the right on several normal faults, which form complicated horst and graben patterns. Sedimentary thicknesses are very uneven, but layers in the graben generally are thicker (Fig. 23) due to the addition of sand during faulting.

At the end of the experiment the graben was not refilled, in order to demonstrate the intersection of the angle of repose of the sand with the fault plane and also with the uppermost key bed. The right side of Figure 23 shows the intersection: the fault dips 72° to the left, and receded

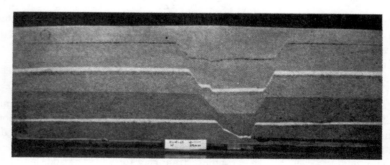

FIG. 20.—Asymmetrical graben in sand. Lower layers were faulted; top layers were filled after faulting. Right fault is steeper and sharp without change in dip; sedimentation in graben is thicker than on "highlands."

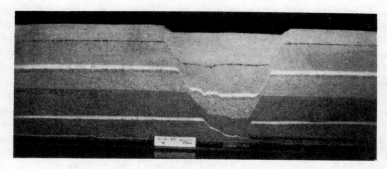

Fig. 21.—Faulting continued. Top layer has been faulted. Fault grows upward without change in dip.

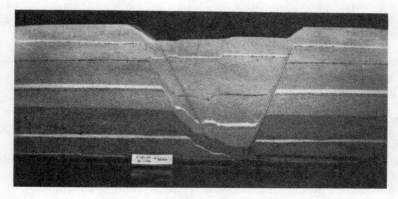

Fig. 22.—Two layers have been added and are faulted. Fault positions at stage of Figure 21 are black lines on glass. Right fault is stationary; left fault migrates toward left with entire block. As graben widens, new faults appear.

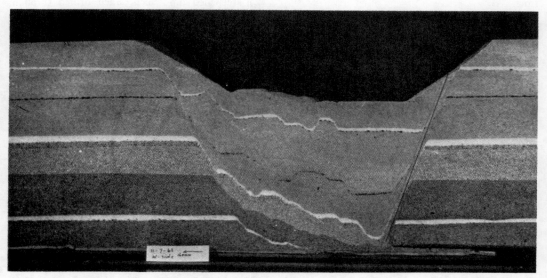

Fig. 23.—Final stage of faulted sand layers. Total displacement is 160 mm. Angle of repose of dry sand is very much less than that of fault dip.

somewhat toward the right as deformation pro-
gressed. The layers are horizontal up to the fault
and break off abruptly at the fault. The sand surface dips about 36° toward the left. The dip depends on packing of the sand, its grain size, and
moisture content. Limestone dust makes vertical
bluffs, as does the upper white layer of white
ground plastic ("microthene") shown in Figure
23.

The intersection of the fault plane with the
angle of repose of the sand is, of course, quite artificial and scarcely is representative of sedimentation in basins receiving sediment. In a sedimentary basin the profile of the sedimentation surface has an angle measured in minutes. Sedimentation continues at a slow rate covering faults as
soon as they reach the surface. Thus, even a fault
with major displacements at depth is not apparent at the surface. The experiment suggests what
the faulting process might be. It also suggests
that the fault continues upward without deflection or change in dip into the younger sediments
as they become buried.

If the entire graben area subsides, thicknesses
should increase abruptly across the master fault
and then may decrease again on the other side.
The largest displacement should be along the
down-to-basin fault. At the same time, the displacement should increase with increasing depth
for those layers which are affected for the longest
time.

In the sand experiment, particle paths (Fig.
24) are very similar to those of the clay experiment. On the right along the master fault, movement parallels the fault and is guided by it. Particles in the stationary right block do not move; in
the left block the movement is horizontal toward
the left. Between these two blocks are areas
where particles move down parallel with the master fault and then turn into a horizontal path
when downward motion is arrested and new
faults open within the graben. The curved paths
are more angular than in clay because sand is not
plastic and cannot bend. The effect is the same
however in all experiments. Points A to D, which
are in one bed, first move along parallel paths.
Then point A turns toward the left but the others
continue downward. This extends distance A-D,
and line A-D begins to dip toward the right. As
this happens more faults appear in order to accommodate the extension. This phenomenon is
also a downbend, even if it is broken somewhat
due to the physical properties of the material
used.

DISCUSSION OF GROWTH-FAULT EXPERIMENT

The experiment shows that the faults continue
upward into new and previously unfaulted layers
which are added above the fault. The dip-slip
therefore grows downward but it also did this in
the clay model where no sediments were added.

The increase in thickness consists of two components: (1) real increase from deposition, and,
(2) steepened dip which, when drilled through,
will show apparent increases in thickness of beds.
It is impossible to appraise the relative impor-

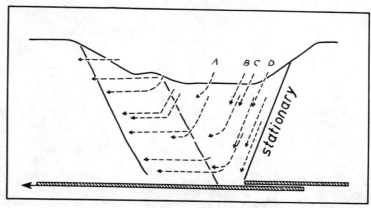

FIG. 24.—Sand-particle paths during deformation (Figs. 21–23). Stationary block is on right with sharp fault
dipping left. Left block moves on lower floor.

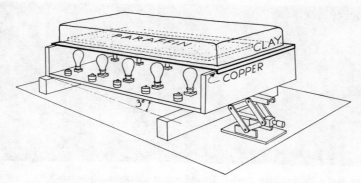

FIG. 25.—Setup for gravitational creep experiment with wet clay (diagrammatic).

tance of these two factors without more information. However, company files must contain abundant information.

In addition to growing faults, the sand-box experiment shows a crude downbend with subsidiary faults. Furthermore, it shows again that the direction of pull has nothing to do with the asymmetry of the resulting structures.

REGIONAL FAULT PATTERN

The preceding discussion shows that down-to-basin faults with reverse drag can be explained rather simply and can be imitated experimentally. It was tempting to try an imitation of the entire semicircular Gulf Coast pattern and graben zones. Many experiments were made until a fair imitation was achieved.

First, a tank was constructed 12 × 36 in. and 1 in. thick with small holes in the top on 1-in. centers. A clay model was built on that surface and the tank inclined 1°. When water pressure was turned on at about 10 lb the entire clay model slid into the sink within moments and essentially in one slab.

The greatest disadvantage of this arrangement is the fact that, when surface tension in clay experiments is relieved by a film of water, shear fractures do not appear but tension fractures do and the model becomes worthless when it breaks up into slabs and irregular blocks.

After much experimentation, the setup of Figure 25 was constructed. It consists of a wooden box above which a copper plate can be heated by rows of light bulbs inside the box. A layer of paraffin on the copper sheet can be modeled to any shape and thickness and the clay placed above it.

The model is tilted a few degrees and the melt-

ing paraffin becomes the lubricant which permits the clay to creep downslope. Fracture patterns form on the surface and the sides of the model. After the desired stage of faulting is reached the lights are turned off and the paraffin hardens. The fractures which cut through to the base of the model are preserved in the paraffin and can be correlated with the surface fractures after the clay is removed.

Figure 26 shows two down-to-basin faults which formed in clay during creep. The angle of inclination of the copper sheet is 3.5°. Reverse dip is toward the right and toward the major faults. Antithetic faults facilitate the downbending as creep proceeds toward the left.

GULF COAST FAULT PATTERN

An attempt was made to duplicate the entire Gulf Coast pattern in one experiment in which creep of a large area was directed away from a stationary land mass of approximately crescent shape. As a pilot experiment, a clay model was built on a metal plate which was pulled. Later the model was built on paraffin and allowed to creep (Figs. 26, 27). Figures 28–32 show the initial experiment, but the pattern is the same in the gravity-induced creep experiment (Fig. 27). Figures 28 and 29 are photographs of the model from above. Figure 28 is illuminated at a low angle from the north (top), and Figure 29, from the south.

For comparison, faults from the Gulf Coast structure map by Murray (1961, Fig. 4.1) are superimposed on the model photograph (Fig. 30). I think that the coincidence is striking. Many well-known structures are represented in the model and some additional ones appear. Some faults dip

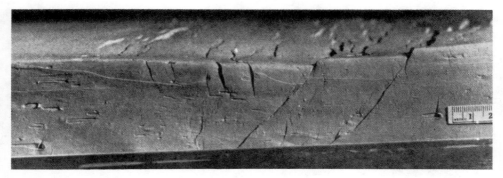

Fig. 26.—Downdip faults in wet clay model. Downbend is clockwise (toward left) with some updip faults visible on surface. Scale in cm.

the wrong way—*i.e.*, a down-to-basin fault in nature is represented by an updip fault in the model. It is nevertheless a fault which indicates a horizontal component of extension. An equivalent of the South Arkansas graben is not present in the model, but if motion had continued it might have appeared. (It would appear late in this model because the clay did not adhere to the metal sheet at the base of the model and therefore lagged behind.) The strike of the faults is the same on the map and in the model, and the east Texas to south Mississippi embayment is represented as well as the down-to-basin fault zones which strike parallel with the coast line.

If the directions of the horizontal components are plotted for both the map and the model the

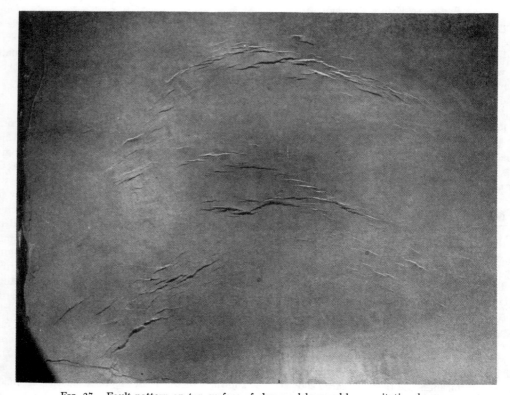

Fig. 27.—Fault pattern on top surface of clay model caused by gravitational creep.

resemblance becomes still more striking. Figure 31 shows the horizontal components added to Murray's map. No attempt was made to indicate the component quantitatively by the length of the arrows. Figure 32 is a similar plot for the model. The difference between Figures 31 and 32 is only in the details. The Mississippi delta area is not shown in the model (Fig. 32), and faults in the model are shown in the Gulf far from shore where they are known to exist.

DISCUSSION OF EXPERIMENTS

The experiments were performed in the following sequence: symmetrical grabens were done many years ago by both H. and E. Cloos. Asymmetry was studied systematically on the suggestion of H. N. Fisk, especially with reference to reverse drag because of the obvious interest of the petroleum geologist in possible closures near down-to-basin faults. The presence of down-to-basin faults clearly proved the existence of a horizontal motion which probably caused both the horizontal and vertical components. Next, sand was added to simulate sedimentation and the asymmetrical faults grew upward into new sediments. Then, an attempt was made to imitate the entire Gulf Coast pattern in one experiment. The results showed similarity with the map pattern in spite of (1) the rather primitive nature of the model, and (2) certain unlikely comparisons: (a) there is no metal sheet being pulled beneath the Gulf Coast and (b) the Gulf Coast fault pattern comprises a huge heterogeneous area with many irregularities in which faults appeared in sequence during a long period of time. However, the experiment shows a horizontal component that results in a pattern similar to nature with rather similar faults. The final effort was aimed at gravity gliding on paraffin which is far less rigid than the metal sheet experiment and therefore may be somewhat more similar to nature.

Even if the method is crude, the resulting patterns are remarkably similar to those which actually exist and suggest that a horizontal component may well be the explanation of the map pattern.

The imitation of the entire Gulf Coast fault pattern in one experiment does not imply that there was only one act of motion during a limited time. The peripheral graben faults appeared early, the faults nearer the coast appeared later, and

some of these are still active. Even in the very short time of that experiment (2–3 hr) not all faults occurred simultaneously but in succession. The first ones appeared after 1 hr, the last ones after 3 hrs.

The sequence in the experiment is similar to that in nature: the marginal faults appear first and the interior ones later, because the clay sheet first tears loose at its margins and then collapses when the fault arcs reach the interior.

The experiments suggest that the regional fracture pattern of the Gulf Coast area was caused by creep of the sedimentary blanket toward the lowest point of the basin. It seems justifiable to conclude that the nearly identical patterns of the experiments and the Gulf Coast are due to similar movements and stress patterns.

The creep mechanism does not require movements of separate basement blocks, or salt intrusions, or flexures and folds for the localization of faults; on the contrary, the mechanism functions better without them. Of course, it cannot be assumed that the basement surface is as smooth as a table top, nor is the Gulf Coast basin fill as homogeneous as the clay in the model. Major obstructions probably do not exist, however, because they would prevent creep. Well-known obstructions, such as the Llano-Burnet uplift, determine the shape of the area which creeps and in turn the pattern of Gulf Coast faults.

Creep begins at a very delicately balanced stage; *i.e.,* when the gravity component barely exceeds the cohesive strength of the sedimentary column, the friction on the glide plane, and the friction on the sides of the basin. Creep is continuous, as shown by contemporaneous sedimentation and faulting, the increase of downward displacements, and the appearance of structures at several stages during the geologic history.

Continuous creep does not necessitate simultaneous faulting of equal order of magnitude at all points. In the model a fault appears here, another one there. Then the displacement dies out and a new fault opens between the first two. Later one of the two faults shows more displacement and the other two seem to become temporarily dormant, *et cetera.*

In the Gulf Coast area, facies changes, thickness changes, transgressions, and innumerable other inhomogeneities will influence the formation of faults and fault zones but the regional fracture

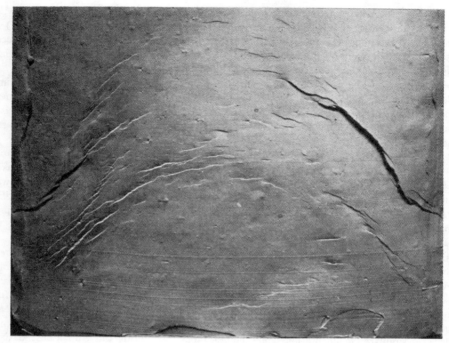

FIG. 28.—Fault pattern on model surface illuminated at low angle from north. Deep shadows are on south-dipping faults; light fault scarps dip north.

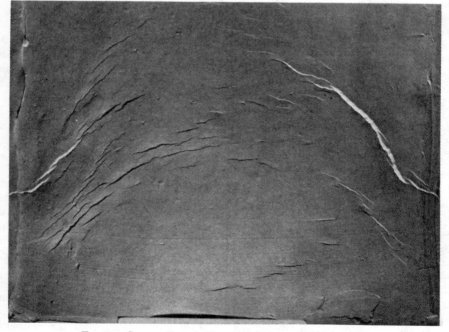

FIG. 29.—Same surface as Figure 28, illuminated from south.

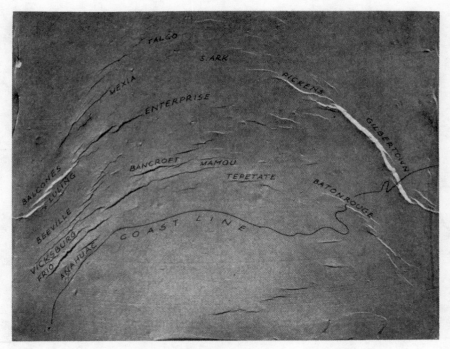

Fig. 30.—Map of Gulf Coast fault pattern superimposed on model (Fig. 29).

pattern transgresses all these and the direction of the horizontal component remains constant.

The peripheral graben zone (Balcones-Mexia-Talco-South Arkansas-Pickens-Gilbertown) follows the "contact" between stationary margins of the Gulf Coast and the central area which creeps away from it. This resembles the familiar "Bergschrund" crevasse where a glacier pulls away from a mountain ice field. It is explained easily as that zone along which the creeping sediments are torn loose from the stationary area. This zone coincides in the Gulf Coast with the approximate subsurface updip edge of the salt, thus suggesting that the salt may well facilitate the creep.

This tear-off zone also is typically not asymmetrical but a more or less symmetrical graben formed between a downdip and an updip fault. Asymmetry is not developed because there is not much creep and the material at the base of the graben is not removed basinward.

The Mexia-Talco *en échelon* grabens are in the position of extension fractures in a zone between a stationary and a moving block as observed at many places on a large and small scale. The existence of such *en échelon* grabens indicates the po-

sition of the boundary between the creeping basin fill and the stationary Llano uplift. On the east side of the basin the boundary of the moving sheet may well flare southeastward and the Pickens-Gilbertown system is therefore not an *en échelon* structure.

INTERFERENCE OF REGIONAL AND LOCAL COMPONENTS

Radial fault patterns are well known in more or less circular domes (Murray, 1961, p. 217, E). The faults are normal faults and prove extension either above an up-arched dome or in its vicinity. In elongate domes or anticlinal updomings, fractures normally are parallel with the long axis of the dome because extension across the dome is greater than extension parallel with the long axes. There are, however, exceptions in true folds in which the limbs have been compressed. This rarely applies in Gulf Coast domes and anticlines. Fractures above circular and elongate domes have been demonstrated experimentally (E. Cloos, 1955, p. 247–253, Pls. 5, 6). In the following section, these fracture patterns are called "local" because they are related to a specific local structure

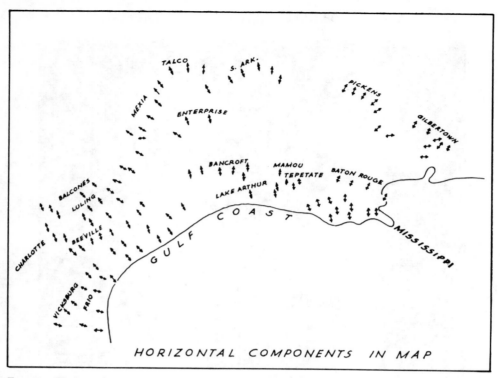

FIG. 31.—Horizontal extension component of Gulf Coast faults, based on Murray (1961, Fig. 21). Arrows show direction; length of arrow meaningless.

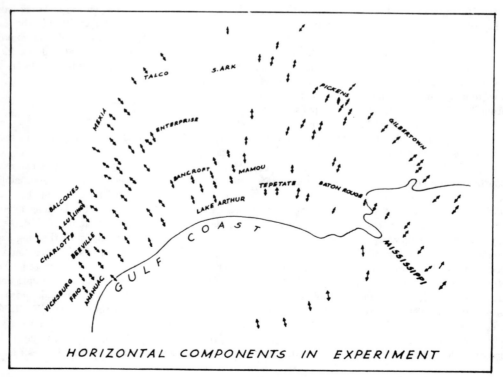

FIG. 32.—Extension components of faults in model, Figures 28–30. Arrows indicate directions, not quantities.

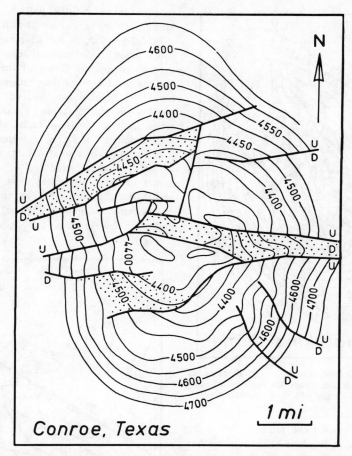

Fᴵɢ. 33.—Conroe field, Montgomery County, Texas (after Carlos, 1953, p. 104). Local components should result in radial faults. Superimposed regional trend is east-west. Compare with Figure 34.

(dome, anticline). The Gulf Coast pattern is called "regional" because it is uniform through large areas. Accordingly, it seems proper to speak of a "local component of horizontal extension," and a "regional component of horizontal extension." This regional component is well shown in Figures 31 and 32.

In elongate anticlinal structures such as salt ridges the two components may coincide and regional faults then parallel the long axis and the local faults. In that case the two components are not distinguished easily.

Where the local and regional components do not coincide, the regional one is triggered by the local one or the local component is subdued by the regional one.

There are many examples of a regional component superimposed on the local component; some of these are shown in Figures 33–37. In each of these examples, the patterns should not be only regional. Obviously the local patterns were suppressed.

Another outstanding example of a dominant regional component is the salt dome at Reitbrook near Hamburg, Germany (Behrmann, 1949, p. 200). This dome is cited frequently as an example of normal faulting above a salt dome. The dome is almost circular or somewhat elongate from northwest (Fig. 38) to southeast. Many normal faults show extension in the direction of the long axis and trend northwest and east-west. Their strikes intersect at angles of 20°–30°. If the

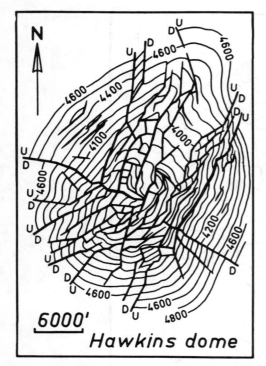

FIG. 34.—Hawkins field, Wood County, Texas (after Wendlandt, 1951). Radial faults are response to local updoming; regional component barely noticeable.

doming alone had caused the faulting, the faults should be radial as at Hawkins (Fig. 34) or other domes.

The lack of evidence at Reitbrook for a local component of extension and the presence of the northeast-trending faults strongly suggest the existence of a regional component oriented northwest-southeast. Behrmann (1949) suggested that the dome had risen above a much broader domed area which provided this regional component. It may not be accidental that long salt ridges northwest of the area also trend northeast and roughly across the dome. The salt ridges and faults above the dome seem to be contradictory unless salt withdrawal toward the northwest provided the regional component. To assume northwest creep analogous to the Gulf Coast creep may seem speculative in the absence of other evidence.

DISCUSSION OF PREVIOUS VIEWS

Gravity sliding, gravity flow, or the "regional component" has been mentioned repeatedly as a possible cause for the pattern of Gulf Coast faults and reverse drag on down-to-basin faults.

Wallace (1952, p. 63) described a "strong regional grain" or a "regional tension developed as a result of general flexing of the north flank of the geosyncline which has continued over a long period of depositional history." He also recognized a "dominant factor in controlling both regional faulting and those faults which cross deep-seated domes" and the fact that "down-to-the-south major faults on structures may be indistinguishable from regional faults." Local faults on domes or other structures have been recognized and called "abnormal." The map by Wallace shows many examples of regional and local faults and their interference. The only real difference

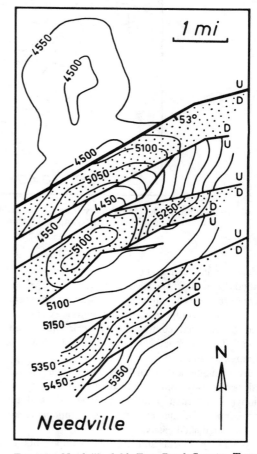

FIG. 35.—Needville field, Fort Bend County, Texas (after Greenman and Gustafson, 1953, p. 140). Regional trend is superimposed on local dome.

between the conclusions of Wallace and this paper is in suggesting that the cause of "regional" faulting is either downwarp (Wallace) or gravity creep (E. Cloos).

Quarles (1953, p. 490) discussed one of the reasons for faulting thus: ". . . the complete column of sediments slips down toward the Gulf on bedding planes and produces a fault where the sliding beds essentially break off the stationary ones." In the subsequent discussion he favored participation of salt ridges in faulting but discussed the "slipping plane hypothesis" (p. 498).

Bornhauser (1958, p. 352) discussed the gravity-flow theory: " . . . folding of sedimentary beds is the result of flowage of incompetent beds

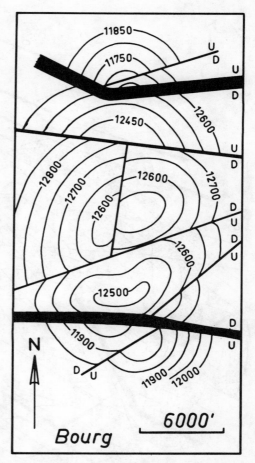

FIG. 37.—Bourg field, Lafourche and Terrebonne Parishes, Louisiana (after DeHart, 1955, Fig. 5). Local north-south anticlinal structure with superimposed regional faults.

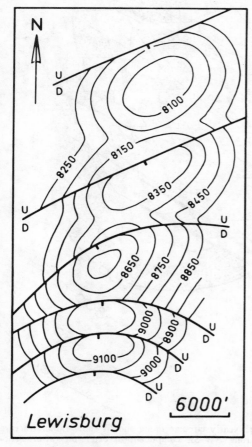

FIG. 36.—Lewisburg field, Acadia and St. Landry Parishes, Louisiana (after Ocamb and Grigg, 1954, Fig. 4). Local north-south-trending anticline with regional faults superimposed.

under load down a sloping surface." He referred to M. King Hubbert (1937, 1945) and suggested evaporites as the locus of possible glide horizons. The flow takes place in waves and, when it takes place, tensional stresses may cause shearing. Bornhauser thought that faulting usually is associated with gravity-flow folding and the process of downfolding provided an explanation for the mechanics involved in this faulting; his reverse drag is linked to folding and folds. My experiments show that folds are not necessary in reverse drag and that gravity flow or creep may be the cause of the fracture pattern. Bornhauser (1958, p. 357) also used the term "master fault"

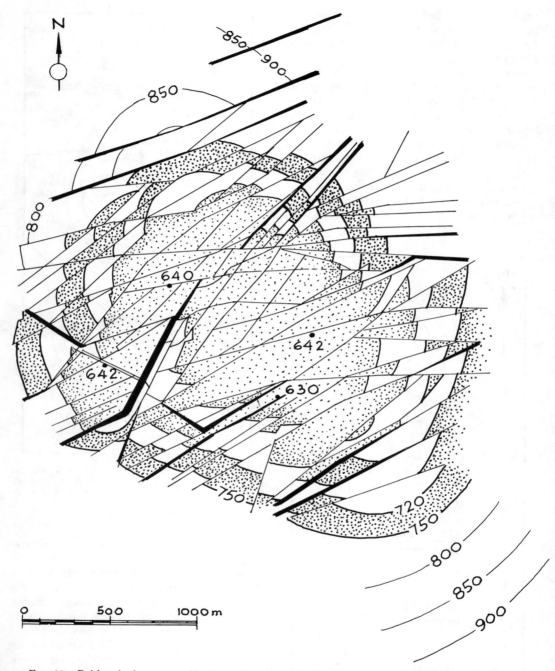

FIG. 38.—Reitbrook dome, near Hamburg, Germany. Nearly circular dome should show radial fracture pattern. "Regional" trend seems to prevail suggesting strong NW-SE component of extension (after Behrmann, 1949, p. 200, Fig. 7).

which seems to be a good term for the fault along which reverse drag is observed.

Hamblin (1965) discussed most recently "reverse drag" on normal faults and showed that the phenomenon is not restricted to Gulf Coast faults. He has mapped reverse drag in the field and gave several examples. Reverse drag is, according to Hamblin, the result of displacement on curved faults.

My experiments confirm the curved-fault principle but show that the curved fault may not be the only explanation. The fault may have to lead into a slip plane which is capable of accommodating the horizontal component.

Hubbert and Rubey (1959) discussed the possibility of lateral movements of rock masses during overthrust faulting and they showed that the presence of high fluid pressures reduce friction at the base of a moving sheet to such an extent that overthrusting becomes likely. They showed also that abnormally high fluid pressures exist in the Gulf Coast area.

Furthermore Rubey and Hubbert (1959, Fig. 9, p. 194) showed a diagrammatic cross section of a zone of abnormal fluid pressure and the beginning of a bedding-plane glide surface on the flank of a geosyncline. The authors stated (p. 194, 195):

Under these conditions, at some depth where the critical relationship between the fluid pressure–overburden ratio and the lateral component of stresses acting on the area is exceeded, a thick section of sedimentary rock would part from its foundation and start to move. The critical relationship between pressure ratio and lateral stresses might be exceeded as the result of unusually rapid deposition, of an increased horizontal compression, or of a slight steepening of the flank of the geosyncline. . . .
In this diagram (Fig. 9) the thrust plane is shown as breaking loose along a normal fault or faults at its rear or upslope margin and sliding downslope into the geosynclinal trough. . . .
The writers do not intend to emphasize unduly the gravitational sliding hypothesis and Figure 9 might just as well have shown no downslope sliding, but a regional compression or horizontal push from the rear. . . . The general principles illustrated in Figure 9 apply without regard to the nature of the forces that cause movement of the thrust plate. The effect of high interstitial fluid pressures is virtually the same— that is, they would greatly facilitate gliding on bedding–plane surfaces—whether the lateral stresses that cause the movement owe their origin to gravitational forces, regional compression, or some other source.

It is quite evident that slip planes, creep, curved faults, and a horizontal component have been thought of by several authors as possible causes of the Gulf Coast fault pattern. The last hypothesis has been worked out most elaborately and seems to be the most convincing.

Conclusions

The experiments presented here demonstrate that the regional Gulf Coast fault pattern may well be the result of gravity creep of the basin fill. Such creep also would provide the mechanism for asymmetries in down-to-basin faulting: curved master faults, reverse drag, antithetic faults, steepening of dips with increasing depth, and even some thickening in the downthrown block.

Creep would provide also the necessary extension for the marginal fault system or tear-off rift at the edge of the basin. If creep began in that zone it would explain its symmetry, *en échelon* graben system, and absence of reverse drag. The assumption of creep is supported strongly by the prevalence of reverse drag on down-to-basin faults and its rarity on updip faults.

Finally, a regional component must cause the suppression of local faults and the prevalence of regional ones in structures that would be expected to show a local fault pattern.

Local patterns are abundant, however, and are without "overprint." Whether regional or local patterns dominate may well be determined by the rate of deformation.

If creep is assumed, no other explanatory mechanisms are needed and the entire Gulf Coast pattern can be explained in rather simple terms by a universal cause.

References Cited

Badgley, Peter C., 1965, Structural and tectonic principles: New York, Harper & Row, 521 p.
Behrmann, R. B., 1949, Geologie und Lagerstätte des Ölfeldes Reitbrook bei Hamburg, *in* Erdöl und Tektonik in Nordwestdeutschland: Hannover-Celle, Amt für Bodenforschung, p. 190–221.
Bornhauser, Max, 1958, Gulf Coast tectonics: Am. Assoc. Petroleum Geologists Bull., v. 42, p. 339–370.
Carlos, P. F., 1953, Conroe field, Montgomery County, Texas, *in* McNaughton, D. A., ed., Guidebook, field trip routes, oil fields, geology: Houston, AAPG-SEPM-SEG, Joint Ann. Mtg., p. 104–109.
Castellot, A. E., and R. P. Caletti, 1958, Exploración de los horizontes profundos del campo José Colomo: Asoc. Mexicana Geólogos Petroleros Bol., v. 10, p. 409–420.
Cloos, Ernst, 1955, Experimental analysis of fracture patterns: Geol. Soc. America Bull., v. 66, p. 241–256.

Cloos, Hans, 1928, Über antithetische Bewegungen: Geol. Rundschau, Bd. 19, p. 246–251.

——— 1930, Zur experimentellen Tektonik: Naturwissenschaften, Jahrg. 18, Hft. 34, p. 741–747.

——— 1931, Zur experimentallen Tektonik, Brüche und Falten: Naturwissenschaften, Jahrg. 19, Hft. 11, p. 242–247.

——— 1936, Einführung in die Geologie: Berlin, Gebrüder Bornträger, 503 p.

——— 1949, Gespräch mit der Erde: 2d ed., Munich (München), R. Piper & Co. Verlag, 389 p.

——— 1953, Conversation with the earth: New York, Alfred A. Knopf, 413 p.

DeHart, B. H., Jr., 1955, The Bourg field area, Lafourche and Terrebonne Parishes, Louisiana: Gulf Coast Assoc. Geol. Socs. Trans., v. 5, p. 113–123.

Greenman, W. E., and E. E. Gustafson, 1953, Needville field, Fort Bend County, Texas, in McNaughton, D. A., ed., Guidebook, field trip routes, oil fields, geology: Houston, AAPG-SEPM-SEG, Joint Ann. Mtg., p. 138–140.

Hamblin, W. K., 1965, Origin of "reverse drag" on the downthrown side of normal faults: Geol. Soc. America Bull., v. 76, p. 1145–1164.

Hardin, Frank R., and George C. Hardin, Jr., 1961, Contemporaneous normal faults of Gulf Coast and their relation to flexures: Am. Assoc. Petroleum Geologists Bull., v. 45, p. 238–248.

Hills, E. Sherbon, 1963, Elements of structural geology: New York, John Wiley & Sons, 483 p.

Hubbert, M. King, 1937, Theory of scale models as applied to the study of geologic structures: Geol. Soc. America Bull., v. 48, p. 1459–1520.

——— 1945, Strength of the earth: Am. Assoc. Petroleum Geologists Bull., v. 29, p. 1630–1653.

——— 1951, Mechanical basis for certain familiar geologic structures: Geol. Soc. America Bull., v. 62, p. 355–372.

——— and William W. Rubey, 1959, Role of fluid pressure in mechanics of overthrust faulting; I. Mechanics of fluid-filled porous solids and its application to overthrust faulting: Geol. Soc. America Bull., v. 70, p. 115–166.

Murray, Grover E., 1961, Geology of the Atlantic and Gulf coastal province of North America: New York, Harper and Bros., 692 p.

Ocamb, R. D., 1961, Growth faults of south Louisiana: Gulf Coast Assoc. Geol. Socs. Trans., v. 11, p. 139–175.

——— and R. P. Grigg, Jr., 1954, The Lewisburg field area, Acadia and St. Landry Parishes, Louisiana: Gulf Coast Assoc. Geol. Socs. Trans., v. 4, p. 183–200.

Quarles, Miller, Jr., 1953, Salt ridge hypothesis on origin of Texas Gulf Coast type of faulting: Am. Assoc. Petroleum Geologists Bull., v. 37, p. 489–508.

Rubey, William W., and M. King Hubbert, 1959, Role of fluid pressure in mechanics of overthrust faulting; II. Overthrust belt in geosynclinal area of western Wyoming in light of fluid-pressure hypothesis: Geol. Soc. America Bull., v. 70, p. 167–206.

Wallace, W. E., 1952, South Louisiana fault trends: Gulf Coast Assoc. Geol. Socs. Trans., v. 2, p. 63–67.

Walthall, Bennie H., and Jack L. Walper, 1967, Peripheral Gulf rifting in northeast Texas: Am. Assoc. Petroleum Geologists Bull., v. 51, p. 102–110.

Wendlandt, E. A., 1951, Hawkins field, Wood County, Texas, in Herald, F. A., ed., Occurrence of oil and gas in northeast Texas: Texas Univ. Bur. Econ. Geology Pub. 5116, p. 153–158.

Copyright 1969 by The American Association of Petroleum Geologists

The American Association of Petroleum Geologists Bulletin
Vol. 53, No. 2 (February, 1969), P. 367-389, 17 Figs., 6 Tables

Structural Analysis of Fractures in Cores from Saticoy Field, Ventura County, California[1]

M. FRIEDMAN[2]

College Station, Texas 77843

Abstract The orientation and relative positions of subsurface faults are inferred from analysis of macro- and microfractures in cores from the Saticoy field, Ventura County, California. The conclusions are reached from study of 1,044 macrofractures (joints) in 4,168 ft of core from 28 wells and from the orientations and relative abundances of microfractures (unhealed fractures in quartz grains) studied in thin sections from cores at several depths in each of six wells. The conclusions are checked by comparison with the known structure of the Saticoy field.

In the main, four macrofracture sets exist along the length of the field. Their "real" rather than observed relative development is assessed upon consideration of the problems of determining the orientation of macrofractures in core samples. In the absence of offset criteria, an interpretation of the fracture data is made solely on geometric grounds and with the geologic knowledge available before the field discovery well was drilled. It is concluded that two of the fracture sets parallel, and therefore define, the orientations of two reverse faults.

The microfractures are shown statistically to be parallel with the macrofractures. In addition, information on the abundance of microfractures is presented for 31 sample locations in six holes that are deviated both toward and away from known faults. The abundance of microfractures increases with proximity to the faults and is essentially independent of the depth of burial.

The strike, dip, and position relative to drill holes of the faults predicted from the fracture data agree with the known subsurface structure.

INTRODUCTION

A structural analysis of fractures in cores from the Saticoy field, Ventura County, California, was undertaken to determine (a) if a reproducible fracture pattern can be measured in cores, (b) if the fractures can be correlated geometrically and genetically to the known structure, (c) if there is a relationship between the abundance of microfractures in the quartz grains of the rocks at different depths and proximities to known faults, and (d) if the

[1] Manuscript received, July 22, 1967; accepted, March 25, 1968. EPR Publication No. 495, Shell Development Company, Exploration and Production Research Division, Houston, Texas. The writer is most grateful to the Shell Oil Company Pacific Coast area staff for use of maps and cross sections, and to Shell Development Company for permission to publish. He particularly thanks John W. Handin for his critical review of the manuscript.

[2] Center for Tectonophysics, Texas A&M University.

fractures could have been used to predict the subsurface structure early in the history of the field. The Saticoy field (Fig. 1) was chosen for this study because the subsurface structure is well known and could serve as a standard for testing the validity of inferences drawn from the fracture data. Moreover, many fractured cores are available that can be oriented from standard subsurface data.

A recent structural interpretation of the field is shown in Figures 1–3 and in several cross sections (Figs. 11–16). The field extends for approximately 7 mi along the east-northeast trend of the Oak Ridge fault and lies about halfway between the towns of Saticoy and Santa Paula. Cores used are from the lower Santa Barbara and upper Pico Formations (late Pliocene), which are composed of turbidites—sandstone, siltstone, and shale. These beds range in strike from N43°E to N72°E, dip from 45°NW to vertical, and are locally overturned. They form the southern limb of the Santa Clara Valley syncline and are bounded on the southeast by the Oak Ridge high-angle reverse fault (*see* Jennings and Troxel, 1954, p. 28–29, for generalized maps and cross sections of the syncline and bounding faults). The Oak Ridge fault is defined as the zone along which Miocene and Oligocene beds (southeast of the fault) have been displaced adjacent to the Pliocene beds. The fault strikes about N60°E and dips 85°–80°SE under the Edwards (EDS) and Santa Paula South (SPS) leases, and 75°–65°SE under the Saticoy lease (SAT); *i.e.*, the dip of the fault decreases eastward. The northerly dip of the Pliocene beds along the length of the field is related to folding and basin growth, and not to drag along the Oak Ridge fault. A second fault north of the Oak Ridge fault also strikes N60°E and dips at a lower angle (Figs. 2, 3, 15). This feature, called the Frontal fault, intersects the Oak Ridge fault at depth and is probably somewhat older. Steeply dipping faults transverse to the trend of the Oak Ridge and Frontal faults are known to exist only near EDS No. 20 and in the vicinity of SPS 56 (between SPS Nos. 39 and 49).

The late Pliocene sandstones are moderately

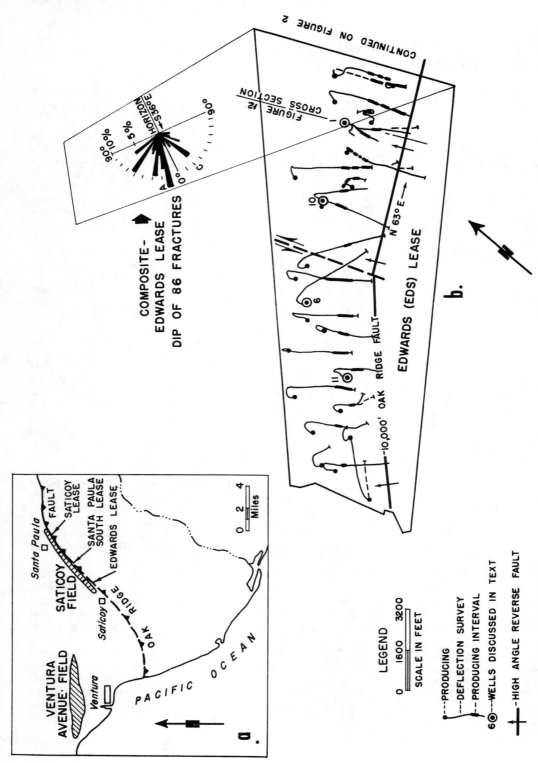

Fig. 1.—Location maps of Saticoy Field (a) and Edwards lease (b), California. Latter shows wells mentioned in text (numbered), trend of Oak Ridge fault at 10,000 ft subsea, and location of cross section in Figure 12. Histogram, a vertical plane, illustrates frequency distribution of dip angles for individual fractures in cores with strikes ±15° of N57°E.

to poorly sorted and contain quartz, feldspar, and clay, and are locally cemented with calcite. They vary from very fine- to very coarse-grained (in places conglomeratic), and have graded bedding. Some beds are nearly unconsolidated, whereas others are indurated by calcite cement. The calcite cement is widespread but not abundant. The calcite is untwinned, and this suggests that it filled pore space that had survived deformation of the beds. Siltstone is conspicuously graded, and in most localities is moderately indurated. Cores of shale usually have parted along natural or artificially induced "planes of weakness" and lack sufficient coherence to permit measurement of fractures or other macroscopic features.

MACROFRACTURES

Method of Study

All available cores from the field were examined in a cursory manner, and those from 28 wells from which cores were particularly abundant were selected for detailed study. This sampling provided good coverage along the length of the field (Figs. 1–3). The cores were oriented from the trace of bedding, dipmeter and hole inclination data, and the known structural geometry. Bedding "tops" were determined from primary sedimentary features in the turbidites. Many cores had to be slabbed and washed to establish the traces of bedding and tops. Recorded data consist of the dip and strike (*in situ*) of the macrofractures (joints), number of fractures, shear displacements along fractures, check on the dip of beds, remarks on lithology and fracture development, and a comment on the reliability of the designated tops. Only those macrofractures sufficiently large and planar to be measured accurately are included in the data. The smallest fracture measured had a surface area of about 1 sq in. In total, 1,044 macrofractures were measured in 4,168 ft of core.

The macrofracture data were analyzed first by plotting normals to fractures in lower hemisphere, equal-area projection in order to assess the overall pattern, and then by identifying in two dimensional plots the existence and average attitude of fracture sets (groups of individual fractures with common orientation). The data were examined further for information on the age of fractures with respect to folding *versus* faulting, changes in fracture pattern along the length of the field, and the relative development of the several fracture sets.

There is a problem in distinguishing natural fractures from those induced by coring, from partings along bedding planes, or from separations that might have occurred as a result of handling the cores. The following operational definitions were adopted: *Unequivocal* natural fractures (Fig. 4) are those that are filled with gouge or vein material or are unfilled but parallel with filled fractures in the same piece of core. These may or may not be slickensided, and may or may not show shear offset. *Very probable* natural fractures are those with slickenside surfaces (Fig. 5). *Probable* natural fractures have clean, fresh planar surfaces and are accompanied by parallel incipient fractures that, in turn, are parallel with unequivocal fractures in nearby cores (Fig. 5b). No other fractures are included in the data presented.

Fracture Data

The fractures exhibit several modes of occurrence in the cores. Forty-three percent occur singly within a given intact piece of core (the average piece is between 3 and 6 in. long), 20 percent occur in sets of 2 parallel fractures, 14 percent in sets of 3, 12 percent in sets of 4, 4 percent in sets of 5, 5 percent in sets of 6, and 3 percent in sets of more than 6. In addition, 36 sets of "pancake" fractures (Fig. 5c) were found, restricted to massive sandstone bodies, spaces from 0.5 to 1.0 in. apart, and commonly characterizing as much as 12 consecutive ft of core. They are inclined at high angles (but rarely normal) to the core axis, are concave up- or downhole, and vary from incipient to well developed. They parallel natural surfaces of mechanical discontinuity present in the rock before coring (Fig. 5c) and probably were opened or propagated by the coring process. The pancake fractures contribute to two of the four major fracture sets (sets A and C). Excluding pancake fractures, 60 percent of the measured fractures occur in siltstone and 40 percent occur in sandstone. This ratio probably reflects the facts that siltstone cores are more readily oriented and only fractures in oriented cores were measured.

The overall macrofracture orientation pattern shows that the normals to the fractures tend to lie in a rather narrow girdle that defines a vertical plane trending north-northwest (Fig. 6a). The perpendicular to the plane of the girdle is horizontal and trends N57°E (Table I). Fractures with normals within the girdle, therefore, have an average strike of N57°E, which is subparallel to the strike of bedding and to that of the Oak Ridge and Frontal faults

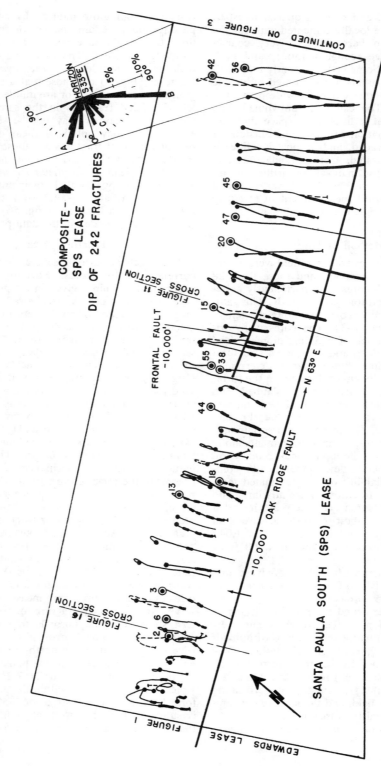

Fig. 2.—Location map for Santa Paula South lease. Legend and explanation similar to that of Figure 1. Location of cross sections on Figures 11 and 16 shown.

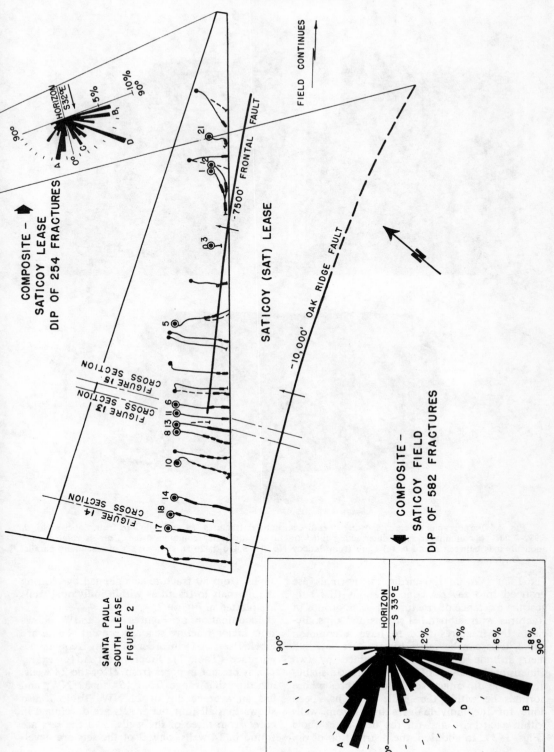

Fig. 3.—Location map for Saticoy lease with legend and explanation similar to that for Figure 1. Composite histogram formed by combining data from histograms for three leases. Location of cross sections on Figures 13, 14, and 15 shown.

Fig. 4.—Photographs of cores showing unequivocal natural fractures. (a and b) Core from Saticoy No. 5, 6,992–7,003 ft, contains array of intersecting filled fractures with nearly common strike. Core is rotated 180° about its axis between a and b. (c) Core from Saticoy No. 14, 9,400–9,420 ft, shows a set of essentially parallel filled fractures.

(N63°E). Within the girdle the normals are grouped into several concentrations that suggest the existence of fracture sets, *i.e.,* groups of fractures with subparallel strikes and dips. Because the fractures tend to have a common strike, tests can be made for fracture sets and their attitude can be better defined by two dimensional analysis of the frequency distribution of the dip angles of those fractures whose normals lie within the girdle. This has been done for the individual fractures from wells within each lease and for the field as a whole (Figs. 1–3). In addition, the distribution of dip angles from 80 fracture sets defined by plotting the normals to fractures within individual wells are plotted in Figure 7.

Examination of Figures 1–3 and 7 shows four major fracture sets are present. These are labeled sets A-D in order of most frequent occurrence (Table I). Fracture set A, for example, is present in cores from 21 of the 28 wells, has dips that range from 15° to 32°NW, and has an average dip of 21°NW. Table I also shows that all four major sets are developed in cores from three of the wells, 3 of the sets are found in 14 wells, only 2 of the sets are devel-

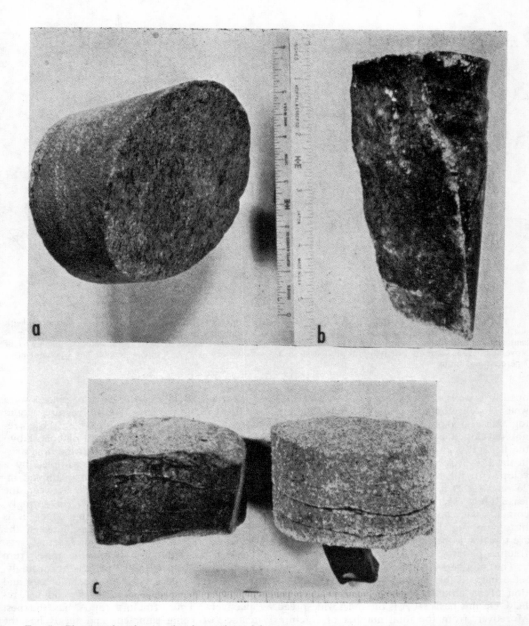

FIG. 5.—Photographs of cores showing variety of fracture types. (a) Probable natural fracture with slicken-sided surface (lineation is in direction of fracture surface dip) in a core from S.P.S. No. 42 O.H., 10,580–10,583 ft. (b) Probable natural fracture (clean, fresh, planar surface) in core from S.P.S. No. 13, 8,300–8,320 ft. (c) Core from S.P.S. No. 2 O.H., 11,451–11,471 ft, shows traces of pancake fractures on siltstone bedding surface. Core on right shows fracture along which separation has occurred, as well as incipient fracture (0.5 in. above most conspicuous fracture).

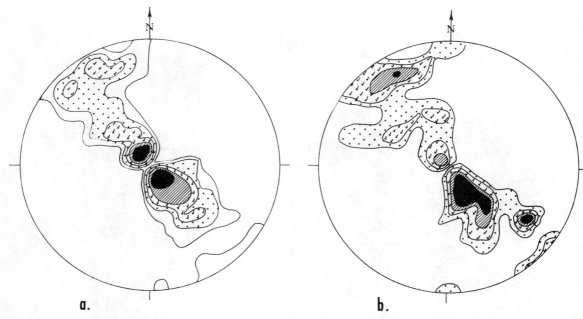

Fig. 6.—(a) Normals to 1,044 macrofractures in cores from 28 wells. Plane of projection is horizontal with north as indicated, lower hemisphere equal-area projection. Contours are at 1, 2, 3, 4, and 6 percent per 1-percent area, 7 percent maximum. (b) Normals to 268 macrofractures in cores from eight wells completed as of November 1956. Diagram oriented and plotted as in a. Contours are at 1, 2, 3, and 4 percent per 1-percent area, 7 percent maximum.

oped in 8 wells, 1 set occurs alone in each of 5 wells; 3 wells contain none of the sets. The histograms also show that the four major sets are not developed uniformly along the length of the field. For example, set A is best developed in the Edwards lease (Fig. 1), set B is best developed in the Santa Paula South lease (Fig. 2), and set D is best developed in the Saticoy lease (Fig. 3). Several fracture sets do not coincide with any of the major sets (e.g., see histogram Fig. 1). Collectively, however, they are too few to be regarded as part of the major fracture pattern.

The number of wells in which a fracture set occurs (Table II, column 2) is one indication of its abundance and distribution in the field. On this basis set A is the most widespread. Moreover, from the total number of fractures that contribute to each of the four sets (Table II, column 3) set A is apparently 1.2 to 2.7 times as numerous as the other sets (Table II, column 4). These values must be interpreted, however, in light of the fact that fractures that intersect the core axis at high angles are preferentially sampled by the core. For example, if a set with fractures 1 ft apart (100 fractures per 100 ft in

a direction normal to the fracture plane) were inclined 90° to the hole axis, all fractures would be cut by the hole and sampled by the corresponding core in a 100-ft length of hole. However, if the set were parallel with the hole axis, at most one fracture would be sampled by a core. The percentage of fractures sampled increases as the sine of the angle between the hole axis and the fracture plane. Accordingly, to correct for this hole-sampling effect, it is necessary to consider the angles between the deviated hole axes and the four fracture sets. The hole deviation for each well ranges from 3° to 33°SE and averages 16°SE. Accordingly, the average angles between the hole axis and fracture sets A-D are 85°, 2°, 58°, and 34°, respectively. Thus, fracture set A has the best chance of being sampled, and set B has the poorest chance. The magnitude of the chance of sampling set A over the other three sets (sine angle between set A and hole axis/sine angle between sets B, C, and D and hole axis, respectively) is listed in Table II, column 5. Comparison of columns 4 and 5 shows that, although the fractures of set A are 1.2 times as numerous as those for set B by actual count, their chances of being sampled are 28 times greater ·than those of set B. By this reasoning set B is the best developed of the four because not only does it occur in 20 wells, but also it is probably much more numerous *in situ* than any of the other sets (Table II, column 6).

MICROFRACTURES

Introduction

The orientation and abundance of microfractures were studied in three mutually perpendicular thin sections cut from each of 31 samples from six wells. The samples were selected on the basis of (a) their location with respect to depth and to the Oak Ridge and Frontal faults, (b) common grain size (150–300 μ), and (c) relative abundance of interstitial material. The study was restricted to unhealed or calcite-filled microfractures developed within individual quartz grains (Fig. 8). This insures that only Pliocene deformation is included in the study, *i.e.*, no relict microfractures related to the previous history of the grains are included. The microfractures were examined with the petrographic microscope and universal stage.

Orientation of Microfractures

The orientation of the microfractures was determined in 10 oriented samples—5 from SPS 2, 3 from SAT 6, and 1 each from SAT 11 and SAT 17. Thin sections were randomly sampled, and the orientation data for each were rotated into a composite diagram to illustrate the microfracture array for each sample (Fig. 9). Within any one grain, only microfractures occurring in groups of two or more parallel individuals were measured. This operational procedure reduces the scatter in the composite diagram that would be overwhelming if every single microfracture were measured and plotted.

Examination of the diagrams in Figure 9 shows that there is a general background scatter in each sample. This is accompanied, however, by a definite tendency for the normals to project in a girdle which defines a vertical plane trending north-northwest across the diagram (*e.g.*, S6, S8, S9, S11, S12, and S24). Most of the highest concentrations (black areas) in the other diagrams also fall within this girdle. The probability that these concentrations occur in a random population is less than 6 in 10,000. The normals within the girdle relate to microfracture planes that have a nearly common strike. This is perhaps more clearly evident in the synoptic diagram (Fig.

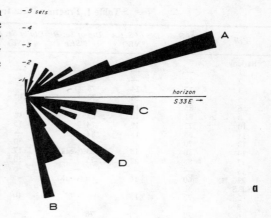

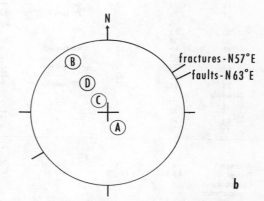

Fig. 7.—Orientation of macrofractures. **(a)** Histogram shows distribution of dip angle of 80 macrofracture sets whose strikes are within 15° of N57°E. **(b)** Lower hemisphere, equal-area projection of average attitude of normals to four sets of macrofractures. Average strikes of macrofractures and of Oak Ridge and Frontal faults are marked at periphery. Normal to Oak Ridge fault varies along length of field, but its average attitude coincides with normal to set B. Similarly normal to Frontal fault coincides with set D.

10), in which only the areas of highest concentration from the individual diagrams are superposed. The average strike of the microfracture planes within the girdle of the synoptic diagram is N56°E. This is in good agreement with the average strike of the macrofractures (N57°E), and with that of the beds, and the Oak Ridge and Frontal faults (N63°E). It is evident that within the north-northwest girdle the maxima correspond to the macrofracture sets A-D (*cf.* Figs. 7 and 10, and *see* Table III).

Fourteen of the 55 maxima in the synoptic

Table I. Fracture Data from Saticoy Field Cores

Well	Dip of Beds (°NW)	Dip of Set A (°NW)	Dip of Set B (°SE)	Dip of Set C (°SE)	Dip of Set D (°SE)	Dip of Misc. Sets (°)	Axis of Girdle (azimuth)	No. of Fractures Measured
Saticoy								
1	—	—	—	—	—	—	—	—
2	72	—	65	—	40	—	N60°E	19
3	77	24	78	25	—	50 NW	N63°E	29
5	85	20	70	(10 to 40)	—	—	N50°E	50
6	85	16	62	25	—	40 NW	N62°E	52+5P*
8	81	16	63	18	—	52 NW	N59°E	35+1P
10	77	17	74	15	—	—	N62°E	26+9P
11	65	18	84	—	40	—	N48°E	25+3P
13	70	20	80	30	—	—	N65°E	25
14	80	15	63 (weak)	17	45	—	N52°E	66+1P
17	85	27	75	10	—	—	N60°E	38+5P
18	—	—	—	—	—	—	—	—
21	60	16	66 (weak)	—	43	45 NW	N55°E	19
S.P.S.								
2	75	15	—	10	42	—	N50°E	28
3	—	—	—	—	—	—	—	—
6	70	30	76	—	34	—	N60°E	62
13	56	22	80	—	—	—	N45°E	52
15	75	32	66	—	37	—	N50°E	44+1P
18	78	—	80 (weak)	12	—	40 NW 59 SE	N53°E	43
20	81	—	77	—	—	—	—	10
36	79	—	75	—	—	39 NW	N55°E	85+1P
38	60	17	66 (weak)	8	39 (weak)	90	N65°E	28+1P
42	45	18	—	12	39 (weak)	70 NW	N58°E	41+1P
44	74	19	75 (weak)	5	—	—	N65°E	20
45 & 47	85	32	—	25	—	—	N62°E	11
55	75	25	59 (weak)	9	—	—	N62°E	23+4P
Edwards								
1	77	18	—	—	—	—	N58°E	80
6	40	—	—	27	—	73 NW	N62°E	22
10	80	—	—	15	—	66 NW	N38°E	23
11	50	22	—	8	—	—	N58°E	47
								Total = 1008+36P
Average	72	21	72	16	40		N57°E	37
Range	40–85	15–32	59–84	5–30	34–45		N38–65°E	
Frequency of occurrence		21	20	18	10			

* P = a set of pancake fractures.

diagram lie outside the north-northwest girdle or are in the girdle but do not coincide with the four major concentrations. These minor maxima tend to fall into groups and are labeled I, II, III, and III' (Fig. 10). Group I contains five maxima from three samples (S20, S40, and S52). The mean position of this group defines a plane that is nearly vertical and oriented N32°W, *i.e.*, essentially normal to the bed-

ding, to the Oak Ridge and Frontal faults, and to the axis of the Santa Clara Valley syncline. Fractures with an orientation normal to a fold axis are common in folded beds (Bonham, 1957; Stearns, 1964; and Price, 1966). (Similarly oriented macrofractures are not present, but they would be difficult to sample by coring.) The percentage of the total number of microfractures contributing to groups II, III,

Table II. Fracture Development

Fracture Set	Frequency of Occurrence (No. of Wells)	Total No. Fractures Contributing to Set*	Ratio: No. of Fractures in Set A to No. in Other Sets	Chance of Sampling Set A over Other Sets	Order of Development
A	21	151	—	—	2
B	20	129	1.2	28	1
C	18	55	2.7	1.2	4
D	10	91	1.7	1.8	3

* Determined from the composite histogram (Fig. 3) by setting limits of dip angle for the fracture sets as follows: Set A, 10°NW–35°NW; set B, 60°SE–85°SE; set C, 5°SE–20°SE; set D, 25°SE–50°SE. Each pancake set was counted as one fracture.

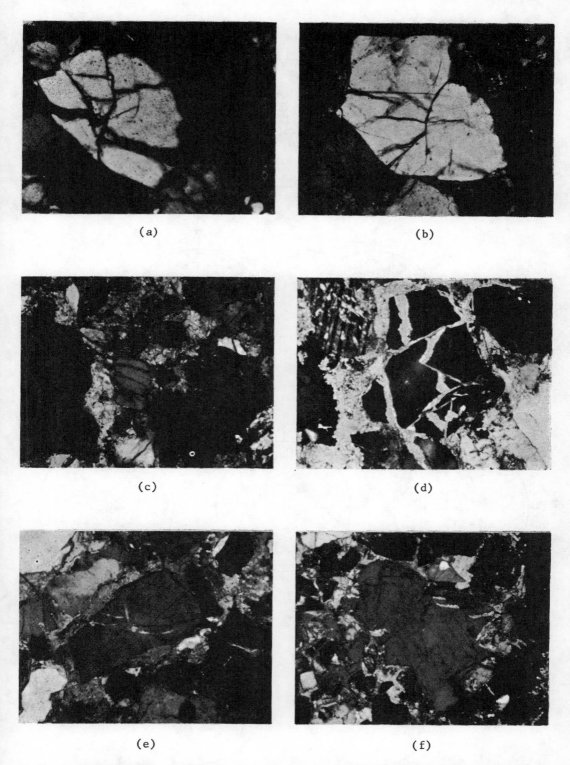

F<small>IG</small>. 8.—Photomicrographs of microfractures. **a, b,** and **c** are fresh unfilled microfractures, **d, e,** and **f** are calcite-cement filled. Note frequent occurrence of two or more nearly parallel microfractures. Crossed nicols. ×100.

161

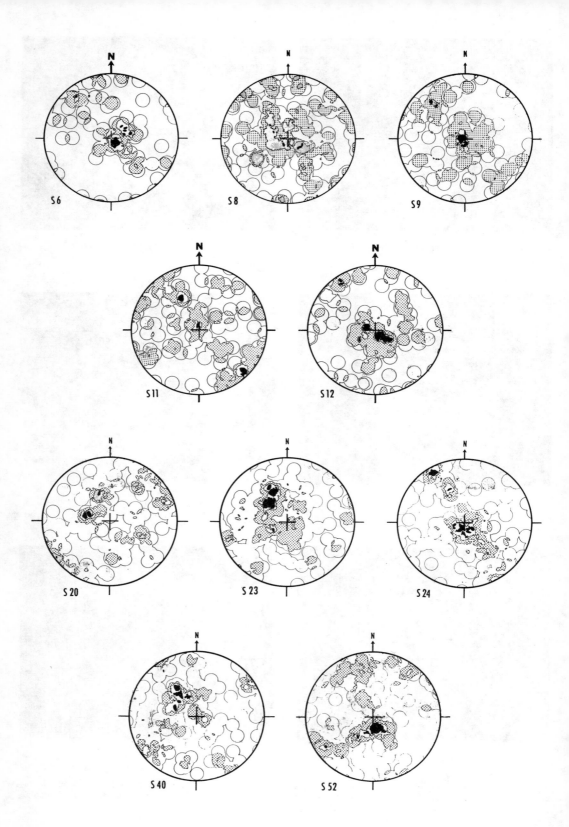

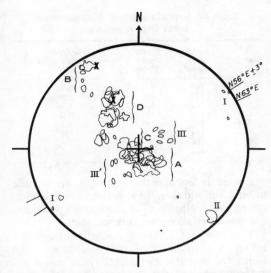

FIG. 10.—Synoptic diagram of 55 statistically significant concentrations of normals to microfractures from individual diagrams (Fig. 9). Normals to the mean attitudes of Oak Ridge and Frontal faults (X) are plotted with strike (N63°E) and mean strike of microfracture planes relating to maxima within girdle. Sets A, B, C, and D are indicated with groups of other maxima I, II, III, and III' discussed in text.

Table III. Correlation Between Microfractures and Macrofractures

Microfractures	Macrofractures
Mean strike of microfractures whose normals fall in the dominant girdle = N56°E.	Mean strike of macrofractures = N57°E.
Dip of set B ranges from 56° to 82° and averages 76°SSE.	Dip of set B ranges from 59° to 84° and averages 72°SSE.
Dip of set D ranges from 28° to 52° and averages 38°SSE.	Dip of set D ranges 34° to 45° and averages 40°SSE.
Dip of set A ranges from 0 to 25° and averages 12°NNW.	Dip of set A ranges from 15° to 32° and averages 21°NNW.
Dip of set C ranges from 0 to 12° and averages 6°SSE.	Dip of set C ranges from 5° to 30° and averages 16°SSE.

and III' is small, and no explanation is offered to account for their occurrence.

The relative development of the major sets of microfractures can be judged directly from examination of the orientation diagrams (Fig. 9). There are no blind spots in the sampling of microfractures when data from three mutually perpendicular thin sections are combined, and no adjustments are necessary because of the coring effect. Set A, the most numerous, appears as a relatively large statistically significant maximum in six of the 10 samples. Set D, the next best developed, appears in half the samples, set B in four, and set C in three. This order of development (A, D, B, C) differs from that found for all the macrofractures (B, A, D, C, Table II). In the Saticoy lease wells, however, the order of development of both micro- and macrofractures is D, A, B, C. Thus the best developed set of microfractures does not necessarily parallel the best developed set of macrofractures.

Comparison of the orientation data (Fig. 9) with the sample locations within the wells (Figs. 11–16) shows that (1) the overall microfracture pattern pervades the domain of sampling, (2) in some cases the pattern is more completely developed nearer the faults, and (3) the best developed set of microfractures is not necessarily parallel with the nearest fault.

FIG. 9.—Composite diagrams showing orientation of normals to sets of microfractures (≥ 2 microfractures/set). Plane of each diagram is horizontal with north as indicated (lower hemisphere, equal-area projection). S6. SPS 2, normals to 77 sets of microfractures in 74 grains. Contours are at 1.3, 2.6, 5.2 and 7.8 percent per 1-percent area, 11.7-percent maximum S8. SPS 2, normals to 145 sets of microfractures in 136 grains. Contours are at 0.7, 1.4, 2.8, and 4.8 percent per 1-percent area, 5.5-percent maximum. S9. SPS 2, normals to 105 sets of microfractures in 99 grains. Contours are at 1, 1.9, 4.8, and 5.7 percent per 1-percent area, 7.6-percent maximum. S11. SPS 2, normals to 113 sets of microfractures in 108 grains. Contours are at 0.9, 1.8, 4.4, and 6.2 percent per 1-percent area, 8-percent maximum. S12. SPS 2, normals to 105 sets of microfractures in 102 grains. Contours are at 1, 1.9, 4.8, 5.7 percent per 1-percent area, 8.6-percent maximum. S20. SAT 6, normals to 128 sets of microfractures in 124 grains. Contours are at 0.8, 2.3, 3.9, and 5.5 percent per 1-percent area, 6.2-percent maximum. S23. SAT 6, normals to 154 sets of microfractures in 149 grains. Contours are at 0.6, 1.9, 3.9, and 5.2 percent per 1-percent area, 8.4-percent maximum. S24. SAT 6, normals to 124 sets of microfractures in 118 grains. Contours are at 0.8, 2.3, 4.0, and 5.6 percent per 1-percent area, 8.1-percent maximum. S40. SAT 11, normals to 132 sets of microfractures in 119 grains. Contours are at 0.8, 2.3, 3.8, and 5.3 percent per 1-percent area, 7.6-percent maximum, S52. SAT 17, normals to 159 sets of microfractures in 142 grains. Contours are at 0.6, 1.9, 3.8, and 5.0 percent per 1-percent area, 8.2-percent maximum.

Abundance of Microfractures

Technique.—The abundance of microfractures was measured to determine whether any systematic relation exists between the microfracturing and proximity to the faults. The abundance is expressed as an index which is subjective. However, if the index is determined by one observer, it can be used to compare the relative abundance of microfractures in a series of samples (Borg *et al.*, 1960; Friedman, 1963; and Handin *et al.*, 1963). The index value for a given sample is determined as follows:

1. In each of three mutually perpendicular thin sections 200 grains are sampled randomly and viewed with the aid of the universal stage.

2. The number of microfractures in each grain is counted, and the grain is classified into one of five arbitrary categories—unfractured, 1–3 fractures, 4–6 fractures, 7–10 fractures, and greater than 10 fractures.

3. The fraction of grains in each category is then calculated.

4. The percentage in the unfractured category is multiplied by 1, that in the 1–3 category by 2, that in the 4–6 category by 3, *etc.* The five products are then summed to give the index value. Thus the index values range from 100 if all grains are unfractured, to 500 if all grains contain more than 10 fractures each.

Several factors influence the microfracture index. Of primary importance is the fact that the index values in experimentally deformed materials increase with increasing strain in the specimens. Superposed on this effect, however, are several other factors. Borg *et al.* (1960) have shown that, in loose, dry aggregates of experimentally deformed St. Peter sand grains, the index increases with increase in grain size for the same experimental conditions. Handin *et al.* (1963) have shown that for the same experimental shortening the index decreases with decreasing effective confining pressure, *i.e.*, the total overburden pressure minus the formation pressure. In addition, it might be expected that the index would decrease with increasing amount of interstitial material, which might tend to cushion the grains from each other. This effect has not been investigated systematically.

In the suite of specimens studied, the formation pressure would be essentially the same throughout the field and would be hydrostatic. However, there could be some increase in fracturing with depth because of the increase in overburden pressure. The factors dealing with grain size and interstitial material were not evaluated independently. Attempts were made to hold variations in these parameters to a minimum by selecting similar sandstone specimens, but small variations are unavoidable.

Table IV. Reproducibility of Microfracture Index Determinations

Sample	Initial Index Value	Repeat Index Value	% Error (Initial—Repeat/Initial) ×100
S51	167	171	2.4
S42	173	168	2.9
S41	162	157	3.1
S30	135	135	0.0
S24	181	173	4.3
S11	161	174	8.1
S6	164	160	2.4
S3	172	173	0.6

The reproducibility of microfracture index values is a function of the consistency with which the operator can identify and count microfractures, and the homogeneity of fracture distribution within the thin sections. In the present study the average reproducibility for eight tests is ±3 percent as listed in Table IV.

Index data.—The microfracture index data are presented on a series of cross sections through SPS 15, EDS 1, SAT 11, 17, 6, and SPS 2 (Figs. 11–16). Samples were chosen to explore possible index trends with respect to proximity to the faults or to depth of burial.

In SPS 15 (Fig. 11) the index values generally increase toward the fault and with increasing depth.

In EDS 1 O. H. and R. D. (Fig. 12) the same trend is observed. In the original hole, O. H., the two values show a marked increase toward the fault; sample S31 is probably fault-zone material.

In SAT 11 (Fig. 13) indices decrease downhole and away from the Oak Ridge fault, with one exception.

In SAT 17 (Fig. 14) the indices first increase with depth and toward the fault and then decrease downhole as the hole gets progressively farther from the fault.

In SAT 6 (Fig. 15) the subsurface structure is complicated by the intersection of the Oak Ridge and Frontal faults. In R. D. 2 the index values, with one exception, increase toward the Frontal fault, then decrease with depth, but increase again near the Oak Ridge fault. The small difference in indices of 150 for S20, located more than 1,200 ft from either fault, and the maximum observed value of 181 (S24) illustrate the subtle nature of the trends.

In SPS 2 (Fig. 16) the trends noted above

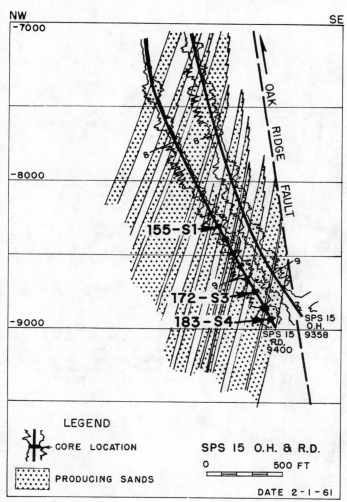

NW

-7000

-8000

155-S1

172-S3

183-S4

-9000

SE

OAK RIDGE FAULT

9

9

9

SPS 15 O.H.

SPS 15 O.H. 9358

SPS 15 R.D. 9400

LEGEND

CORE LOCATION

PRODUCING SANDS

SPS 15 O.H. & R.D.

0   500 FT

DATE 2-1-61

FIG. 11.—Cross section through SPS 15 shows microfracture index data for three samples (S1, S3, and S4). O.H. means original hole; R.D. means redrill.

are not followed. In both the O.H. and R.D. 3, where nearly constant horizontal distances to the mapped position of the Oak Ridge fault were maintained, the index tends to decrease downhole. Of special interest is the fact that, with the exception of the index for S11, there is no significant change from S9 to S13 (from $-7,200$ to $-10,200$ ft).

DISCUSSION AND INTERPRETATION OF DATA

Microfracture Abundance

The microfracture index trends in SPS 15, EDS 1, SAT 11, 17, and 6 indicate that the amount of fracturing increases with proximity to the faults. This consistency suggests that the small lithologic differences among samples are of negligible importance. That this effect also overshadows the influence of depth of burial is apparent from comparison of the index values in SPS 2 (Fig. 16) with those for the other wells. In SPS 2, with the exception of sample S11, there is no significant difference in the values in a 3,000-ft interval. In SPS 15, on the other hand, there is a marked increase in the values within a depth interval of 700 ft, but the horizontal distance to the Oak Ridge fault also decreases. Moreover, in SAT 11 or SAT 17 an abrupt decrease in values takes place within a 1,000-ft depth interval, where the samples increase in horizontal distance from the fault. Accordingly, the influence of depth of burial

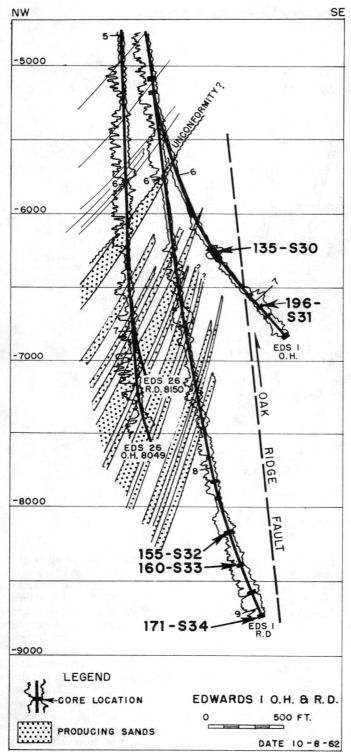

NW

SE

-5000

UNCONFORMITY?

5

6   6   6

-6000

135-S30

196-
S31

-7000

EDS 1
O.H.

EDS 26
R.D. 8150

OAK   RIDGE   FAULT

EDS 26
O.H. 8049

8

-8000

155-S32
160-S33

9

171-S34

EDS 1
R.D

-9000

LEGEND

←CORE LOCATION

PRODUCING SANDS

EDWARDS 1 O.H. & R.D.

0   500 FT.

DATE 10-8-62

Fig. 12.—Cross section through EDS 1 shows microfracture index data for five samples (S30-S34). O.H. means original hole; R.D. means redrill. Location shown on Figure 1.

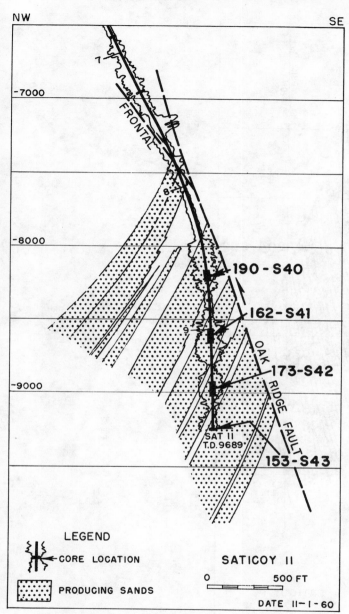

FIG. 13.—Cross section through SAT 11 shows microfracture index data for four samples (S40-S43). Location shown on Figure 3.

and lithologic parameters is at best secondary in contrast to the consistent trends with respect to the faults.

The microfracture abundance data are inconsistent at several stations. Reversals in the overall trend may be noted in samples S41 and S43 in SAT 11 (Fig. 13), samples S21 and S22 in SAT 6 (Fig. 15), and samples S6 and S8 in SPS 2 (Fig. 16). In each of these cases, how-ever, the overall trends are established by the accompanying samples. A fourth inconsistency occurs for S30 in EDS 1 O.H. (Fig. 12), where the index value is low in view of the close proximity of this sample to the Oak Ridge fault. That this value is real is indicated by its reproducibility (Table IV). The essentially similar index values at samples S9, S10, S12, and S13 in SPS 2 (Fig. 16) are puzzling

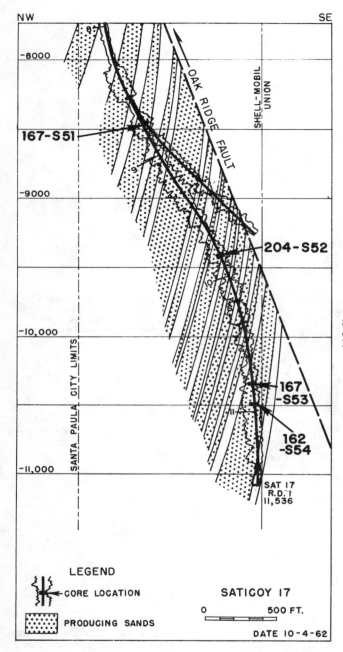

FIG. 14.—Cross section through SAT 17 shows microfracture index data for four samples (S51-S54). Location shown on Figure 3.

in that the first two samples are located within 250 ft of the mapped position of the Oak Ridge fault, whereas the last two are more than 1,600 ft away. That the fracturing in the most distant pair relates to the Oak Ridge and Frontal faults is clearly evident from the microfrac- ture orientation data (compare diagrams for S12 with those for S6 or S8, Fig. 9). The writer can offer no explanation for these index values that is compatible with all the available data. One might speculate, however, that a fault exists below S12 and S13.

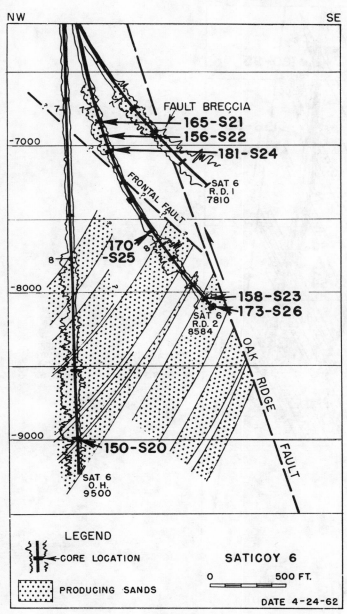

NW

SE

FAULT BRECCIA
165-S21
156-S22
181-S24

SAT 6
R.D. 1
7810

FRONTAL FAULT

**170
-S25**

158-S23
173-S26

SAT 6
R.D. 2
8584

OAK
RIDGE
FAULT

-7000

-8000

-9000

150-S20

SAT 6
O.H.
9500

LEGEND

←CORE LOCATION

PRODUCING SANDS

SATICOY 6

0   500 FT.

DATE 4-24-62

Fig. 15.—Cross section through SAT 6 shows microfracture index data for seven samples (S20-S26). Location shown on Figure 3.

Relative Age of the Fractures

An important key to the interpretation of the fracture data is the relative age of the fractures with respect to the folding of the beds. Should the interpretation be based on (a) the present attitude of the fractures (if they occurred after the folding), (b) a restored geometry obtained by unfolding the beds (if fracturing occurred early in the history of folding), or (c) some intermediate situation?

Three lines of evidence suggest that, in the main, the fractures were formed after the beds were folded. First, the normals to the fractures are grouped statistically into concentrations and are not mingled, as might be expected if the fractures developed at different stages during

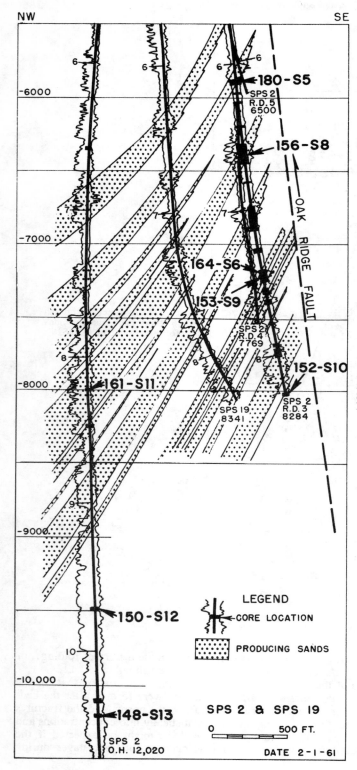

Fig. 16.—Cross section through SPS 2 and 19 shows microfracture index data for eight samples (S5, S6, S8-S13). Location shown on Figure 2.

LEGEND

CORE LOCATION

PRODUCING SANDS

SPS 2 & SPS 19

0   500 FT.

DATE 2-1-61

folding, or if some of the fractures developed late and some early (Figs. 6, 7). Second, the relative simplicity of the fracture pattern itself suggests that the fractures were formed subsequent to the folding. Fracture patterns on well-exposed surface folds are complex and usually involve several shear and extension fracture arrays that apparently are genetically related to bending moments (Stearns, 1964; Price, 1966). Such fracture arrays are not a part of the major four-set pattern obtained here. Third, the abundance of microfractures is clearly related to the faulting, and since the microfractures and macrofractures are parallel, at least statistically, all the fractures apparently are related to faulting rather than folding. It is concluded, therefore, that the fractures now present in the rocks should be interpreted as having formed after the folding, *i.e.*, after they acquired their present attitude.

Interpretation

Ideally, it would be desirable to identify the fractures forming the four major sets as either shear or extension features. However, not enough instances of shear offset were observed to do this directly, and reasoning from the fracture pattern alone leads to ambiguity. It may be assumed that each of the fracture sets is composed of shear fractures, and that faults might exist in the subsurface parallel with each of them. That is, set B might be parallel to a fault steeply dipping southeast, set D might define a fault inclined about 40°SE, and sets A and C might be parallel with faults inclined at less than 30° northwest and southeast, respectively. No assumption is made on sense of shear along these faults. Since any hole drilled in the field would be likely to intersect the low angle faults (A and C), and since none was found by early drilling, the existence of these faults is ruled out. This leaves the prediction from the fracture data alone that at least two faults striking N57°E and dipping an average of 72°SE and 40°SE respectively, might exist in the field.

Before SPS No. 2 (the field discovery well) was drilled in September 1954, the general east-northeast trend and high-angle, reverse nature of the Oak Ridge fault had been established, but its dip at depth was unknown. Two earlier wells, General Petroleum SPS No. 1 and Shell SPN (North) No. 1, were drilled to greater than 9,000 ft subsea and did not intersect the fault. From these data and a seismic line across the structure the possibility of a

northerly dip seemed small, and the Oak Ridge fault was thought to be nearly vertical to at least 9,000 ft subsea. The dip of the fault remained conjectural until November 1956, when (a) data from SAT Nos. 1–4 indicated that the Oak Ridge fault dips 60°SE below 6,000 ft and that there might be a fault (now called the Frontal fault) striking generally parallel with the Oak Ridge in the area of these wells, (b) a dip of 70°SE was recorded in Shell Harvey No. 1 (east of Saticoy lease), and (c) steeper but still southeasterly dips were suspected in the Santa Paula South and Edwards leases.

With only the information on hand before the discovery well was drilled, therefore, one could have concluded that fracture set B was parallel with the Oak Ridge fault and set D probably defined a second fault. Moreover, because set B is better developed, it might also have been concluded that the second fault was subordinate to the Oak Ridge fault. These conclusions check with the nature of the Frontal fault.

A speculation on the sense of shear for set D (and the Frontal fault) can be made once set B is regarded as a shear fracture set parallel with the Oak Ridge reverse fault. The average angle between sets B and C is 56° and that between sets A and D is 61° (Table I). These angular relations suggest that set C is also a set of shear fractures conjugate to set B and that sets A and D form a system of conjugate shear fractures. This viewpoint is supported by the variation in dip angles for sets A and D where they occur together in cores from the same wells (Fig. 17). The southeast dip of set D decreases as the northwest dip of set A increases so as to maintain an average dihedral angle of 61°. It follows that a reverse sense of shear should be assigned to set D, and the inferred parallel fault would then be a reverse fault. The correlation between the fracture data and the known fault pattern for the Saticoy field is listed in Table V.

Historical Approach

An examination of the fracture data from the eight wells completed up to November 1956 (Table VI) shows that at least two occurrences of each of the four major fracture sets delineated by the study of all the data are present in the cores. Moreover, the normals to the individual fractures have the same orientation pattern as those from all the wells combined (Fig. 6b). It could have been concluded that the strike of the faults averages N62°E in the

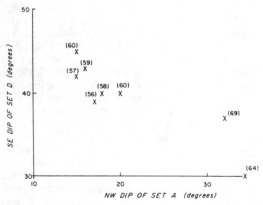

FIG. 17.—Relations between dip of sets A and D in eight cases. Values in parentheses represent angle (°) between fracture sets; average angle for eight cases is 60°.

region of the Saticoy lease, N52°E in the Santa Paula South lease, and N58°E in the vicinity of EDS No. 1. Moreover, it could have been predicted that the Oak Ridge fault would dip an average of 71°SE in the Saticoy lease and 78°SE in the Santa Paula South lease. These conclusions are in agreement with the total fracture data and with the known subsurface structure.

SUMMARY AND CONCLUSIONS

Cores from the Saticoy field, Ventura County, California, contain both macro- and microfracture patterns consisting of four major sets (designated A-D). Reasoning from the fracture data and the structural information available before the field was discovered, it can be concluded that set B is parallel with the Oak Ridge fault. Shallow dipping faults parallel with sets A and C are excluded because any such faults probably would have been detected by any well drilled. Set D is the only other fracture set that could parallel a fault that might not have been detected by early drilling. The correlation between this interpretation and

Table V. Correlation Between Fractures and Known Structure

Known Structure	Fracture Data
1. Strike of Oak Ridge fault is N63°E (Figs. 1–3).	1. Strike of fractures averages N57°E (Table I).
2. Angles of dip on the Oak Ridge fault below 6,000 ft range from 60° to 85°SE and average 75°SE.	2. Dip of set B ranges from 59° to 84°SE and averages 72°SE (Table I).
3. Dip of Oak Ridge fault steepens along trend westward from 60°SE in the eastern part of the Saticoy lease to 84°SE near S.P.S. Nos. 2 and 6 in the Santa Paula South lease. Average dip is 85° in the Edwards lease.	3. The least-square line for a plot of the dip of set B *versus* distance along the fault trend shows the same steepening, 69° to 75°SE.
4. Dip of Oak Ridge fault is constant with depth to at least 10,000 ft subsea.	4. Data inconclusive.
5. Frontal faults are known in the Saticoy lease (dip=45° SE in SAT No. 6, Fig. 15) and in the Santa Paula South lease (Fig. 2).	5. Set D is regarded as parallel with the Frontal fault because (a) the dip of set D agrees with the dip of the Saticoy lease Frontal fault; (b) set D is best developed in the Saticoy lease cores (Fig. 3), where the Frontal fault is perhaps also best developed; (c) set D occurs in cores from SAT Nos. 2, 5, 10, 14, and 21 and from S.P.S. Nos. 2, 15, 18, and 38. Frontal faults are known to occur in all except S.P.S. Nos. 2 and 18.
6. Cross-faults are known only near Edwards No. 20 and S.P.S. No. 56.	6. Fracture sets possibly related to cross-faults are not developed.

the known structure is excellent. In addition, macrofracture data from eight wells completed by November 1956 were analyzed separately to determine whether the four-set fracture pattern could have been detected early in the development of the field. These data yield the same information as the combined data from the 28 wells.

The abundance of microfractures increases with proximity to the Oak Ridge and Frontal

Table VI. Review of Data from Wells Completed as of November 1956

Well	Date Survey Issued	Fracture Sets and Dip	Strike of Fractures
SAT. 1	8/56	None	
SAT. 2	10/56	B (65°SE), D (40°SE)	N60°E
SAT. 3	11/56	A (24°NW), B (78°SE), C (25°SE)	N63°E
S.P.S. 2	3/55	A (15°NW), C (10°SE), D (42°SE)	N50°E
S.P.S. 3	7/55	None	
S.P.S. 6	6/56	A (30°NW), B (76°SE), C (34°SE)	N60°E
S.P.S. 13	105/6	A (22°NW), B (80°SE)	N45°E
EDS. 1	3/56	A (18°NW)	N58°E

faults. This trend clearly overshadows the influences of increasing effective pressure with depth and of the small lithologic differences that exist among the samples. The abundance trends, considered in light of sample location in the deviated holes, permit an estimate of the relative position of the Oak Ridge fault with respect to the drill holes. Moreover, the microfracture abundance and orientation data in SAT 6 suggest independently that the Frontal fault has been pierced by that drill hole.

In retrospect, had a petrofabric study of the cores been made early in the history of the field, there would have been sufficient information to infer the nature of the subsurface fault pattern. With this information the field could have been developed more efficiently. It would have been suspected that the Frontal fault extended along the entire length of the field; and the reservoirs of the frontal and intermediate blocks, which are separated by the Frontal fault, could have been exploited simultaneously by suitably deviated wells.

REFERENCES CITED

Bonham, L. C., 1957, Structural petrology of the Pico anticline, Los Angeles County, California: Jour. Sed. Petrology, v. 27, p. 251–264.

Borg, I. Y., et al., 1960, Experimental deformation of St. Peter sand—a study of cataclastic flow, p. 133–191, in D. T. Griggs, ed., Rock deformation: Geol. Soc. America Mem. 79, 382 p.

Friedman, M., 1963, Petrofabric analysis of experimentally deformed calcite-cemented sandstones: Jour. Geology, v. 71, p. 12–37.

Handin, J., et al., 1963, Experimental deformation of sedimentary rocks under confining pressure—pore pressure tests: Am. Assoc. Petroleum Geologists Bull., v. 47, p. 717–755.

Jennings, C. W., and B. W. Troxel, 1954, Ventura basin, geologic guide no. 2, in R. H. Jahns, ed., Geology of southern California: California Div. Mines and Geology Bull. 170, 63 p.

Price, N. J., 1966, Fault and joint development in brittle and semi-brittle rocks: Oxford, Pergamon Press, 172 p.

Stearns, D. W., 1964, Macrofracture patterns on Teton anticline, northwestern Montana (abs.): Am. Geophys. Union Trans., v. 45, p. 107.

Reservoirs in Fractured Rock[1]

DAVID W. STEARNS and MELVIN FRIEDMAN

Center for Tectonophysics, Texas A&M University, College Station, Texas 77843

Abstract In recent years three developments which have evolved more or less independently, when related, may be of value to the petroleum industry. First is the recognition, through normal oil field development, that fractures are significant to both reservoir capacity and performance. Second is the fact that controlled laboratory experiments have produced, in increasing quality and quantity, empirical data on rupture in sedimentary rocks. These data have been segregated to demonstrate the individual control on rupture of several important parameters: rock type, depth of burial, pore pressure, and temperature. The third development consists of the discovery of new methods to recognize, evaluate, use, and, in some cases, see fractures in the subsurface. This discussion of these three developments may help geologists and engineers to find new approaches to exploration and exploitation of fractured reservoirs. Reservoir and production engineers presently make the greatest use of fracture data, but geologists should find this information useful in exploration for oil and gas trapped in subsurface fractures. Except in the search for extensions to proved fracture reservoirs, there is in the literature a paucity of clear-cut examples of the use of fracture porosity data in advance of drilling. For this reason, several speculative exploration methods discussed herein implement mapping of fracture facies as well as stratigraphic facies.

INTRODUCTION

Scope

In the past 15 to 20 years the petroleum industry has shown an increasing interest in fractured reservoirs. During this same period, laboratory results have provided an empirical evaluation of the factors that influence fracturing in rocks, and advances in methods for studying fractures in the subsurface have been numerous. Nevertheless, few fractured reservoirs have been discovered as a result of exploration specifically for such traps. The succeeding review of pertinent data on rupture in rocks and of selected methods for studying fractures in reservoirs may lead to exploration specifically for

[1] Manuscript received, December 19, 1969.

We especially thank John Handin for his thoughtful reading and constructive criticism of the manuscript and Robert Berg for his many helpful suggestions.

The fracture-orienting device described in this paper was designed while we were employed by the Exploration and Production Research Laboratories of Shell Development Company, Houston, Texas. We are grateful to Shell Development Company for their permission to publish information on this device. We also wish to thank Shell Development for the use of permeability data from cores.

fractured reservoirs or, at least, to their recognition early in the development of a field. Knowledge of the fracture control in a given reservoir not only aids in the primary recovery of hydrocarbons, but also guides in the design of secondary recovery programs.

Basic Definitions and Concepts

A material is said to rupture when it loses cohesion along a more or less planar surface, separates into discrete parts, and is no longer able to support a stress difference. For most materials, this definition is unambiguous, but, for sedimentary rocks in the earth's crust, the concept is not quite so clear. Scale and boundary conditions become important factors. For example, a single rock layer interbedded with several other lithologic types may contain many "fractures"; however, it cannot be separated from its geologic environment, and the entire interbedded sequence may not be "fractured" by our definition. Without the constraints by the surrounding beds, the fractured layer would not be able to support a stress difference; but, as a part of the entire layered mass which has not lost cohesion, it still may be able to carry a part of the load and, indeed, even to rupture further. It is essential in the consideration of fractured reservoirs to differentiate between the behavior of the individual fractured layer and the bulk behavior of the entire sedimentary section.

A discussion of fractures and stratigraphic traps in a single volume might be questioned, because the common cause of fracturing is tectonism. Although fractures are associated with structure, a fractured reservoir might be considered as a form of stratigraphic trap—because lithology controls, to a large extent, the formation of fractures, and a stratigraphic trap is ". . . a type of trap which results from a variation in lithology of the reservoir rock . . . ." (AGI, 1957). Since most rocks are fractured in the subsurface, fractures influence the productivity of nearly all reservoirs.

Fractures play a multiple role in reservoir alteration. They can change the porosity or the permeability, or both. If the fractures are filled with secondary minerals, they may inhibit fluid flow. However, even in rocks of low matrix porosity, fractures may increase the pore volume so that hydrocarbons can be recovered profitably. In other rocks the fractures may not add substantially to the total pore volume, but they may connect previously isolated spaces. Moreover, fractures actually may contribute to reservoir performance by both increasing the porosity and connecting isolated matrix porosity to the well bore.

Good general reviews of fractured reservoirs and their economic significance were given by Hubbert and Willis (1955), Smekhov (1963), and Drummond (1964). The typical fractured reservoir is (1) developed in brittle rock with low intergranular porosity; (2) characterized by high permeability (*e.g.,* up to 35 darcys; Regan, 1953) and low bulk porosity (*e.g.,* <6 percent; Hubbert and Willis, 1955); (3) described in terms of two distinct elements, fractures and intact rock, each of which has its own characteristic porosity and permeability; and (4) recognized initially by loss of drilling fluid, data derived by use of certain logging techniques, production many times that expected from intergranular porosity and permeability, pressure interference with offset wells as much as 50 mi (80 km) distant (O'Brien, 1953), erratic productivity from different wells in the field, and the general enhancement of production by artificial stimulation. In the past, fractured reservoirs usually have been discovered by accident.

Significant differences exist among fractured reservoirs. 1. The pores of the host rock may or may not contain hydrocarbons (Walters, 1953; Doleschall *et al.,* 1967). 2. The reservoir potential may or may not be indicated by open-hole drill-stem tests (Drummond, 1964; Harp, 1966). 3. Under certain conditions the fluids segregate into zones of high oil and high gas saturation that result in the increase of gas-oil ratios early in the depletion history (Pirson, 1953). 4. For some reservoirs, it is possible, on the basis of buildup and/or drawdown plots (Warren and Root, 1963), to distinguish between fractured reservoirs and those involving only intergranular porosity; for others it is not (Odeh, 1964). 5. The fractures creating the reservoir may be genetically related to local faults and folds (Braunstein, 1953; O'Brien, 1953; Daniel, 1954; Kafka and Kirkbride, 1960; Ogle, 1961; Park, 1961; Pohly, 1962; Ells, 1962; Malenfer and Tillous, 1963; Martin, 1963; Pirson, 1967), may be part of a regional system (Hunter and Young, 1953; Walters,

1953; Wilkinson, 1953; Elkins, 1953; Durham, 1962; Pampe, 1963; Ward, 1965), or they may be related to both (Regan, 1953; Daniel, 1954).

DISCUSSION OF FRACTURE PROCESS
Fracture Phenomena

There are two types of fractures. Shear fractures involve movement parallel with the plane, but there is no perpendicular movement. In extension fractures, the walls move apart. No adequate theory explains either the formation or the propagation of either type of fracture. However, the geometric relations of the fractures with respect to the causative stress state have been well established empirically in the laboratory. For any triaxial stress state[2] there are two potential shear-fracture orientations and one potential extension-fracture orientation. The two shear-fracture planes form with a dihedral angle of about 60°. The axis of greatest principal stress bisects this acute angle; the axis of least principal stress bisects the obtuse angle; and the intersection of the two planes is parallel with the intermediate principal stress axis. The extension fracture is parallel with the plane containing the greatest and intermediate principal stress axes and normal to the least principal stress axis.

The geometry of shear fractures is in accord with the Coulomb (1776) criterion, which states that shear fracturing is controlled not only by the initial shear strength of the material but also by its internal friction. Coulomb suggested that shear fractures would occur, not along planes of maximum shear stress (45° to σ_1), but, because of internal friction, at some angle of less than 45°. The departure from 45° is a function of internal friction and is a property of the material. Specifically, shear fractures develop at $\pm 45° \mp \phi/2$, where ϕ is the angle of internal friction. Though this criterion is nearly 200 years old, its significance has been appreciated by geologists only recently. Experimental work (Handin, 1966) and field work (Stearns, 1967) substantiate that, even in the absence of good mechanistic theory, the Coulomb criterion can be used to predict the geometric relation of shear fractures to the axes of the three principal stresses in the rock at the

[2] In this paper the greatest principal stress axis will be indicated by σ_1, the intermediate principal stress axis by σ_2, and the least principal stress axis by σ_3. Therefore, a triaxial stress state is one in which $\sigma_1 > \sigma_2 > \sigma_3$ compressive stresses are considered positive.

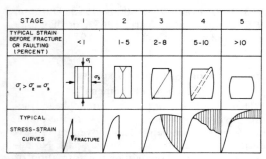

STAGE	1	2	3	4	5
TYPICAL STRAIN BEFORE FRACTURE OR FAULTING (PERCENT)	<1	1-5	2-8	5-10	>10
$\sigma_1 > \sigma_2 = \sigma_3$					
TYPICAL STRESS-STRAIN CURVES					

FIG. 1—Generalized spectrum of deformation characteristics of rocks as measured and observed in laboratory (after Griggs and Handin, 1960).

time of rupture. The position of the extension fracture, though not included in the Coulomb criterion, can be predicted from the results of this same experimental and field work.

Terminology

Several useful descriptive adjectives can be used to modify the term "fracture." In this paper the term *conjugate* implies fractures that are related by a common origin and that form under a single state of stress. A complete conjugate pattern contains a left-lateral shear, a right-lateral shear, and an extension fracture. Because conjugate fractures arise from the same state of stress, the whole pattern of the three potential fractures relative to one another can be determined at any point where the orientation of only one of the fractures is known. Conjugate shear fractures form in most rock materials in such a way that the angle between them is about 60°.

The term *orthogonal* will imply only that fractures are normal to one another. Because of the restriction to a single stress state placed on conjugate fractures, orthogonal fractures cannot also be conjugate. This is not to say that orthogonal fractures cannot have a common geologic origin; but, because they are normal to one another, whether shear or extension fractures, they cannot be caused from a single stress state, as far as is now known.

Regional fractures are those that pervade a wide area, such as the Colorado Plateau, and apparently are unrelated to local structures. Conversely, structure-related fractures are directly associated with an individual structural feature. These fractures also may be found throughout vast areas, but only where similar local structural features also pervade that area.

Fracture number refers to the average number of parallel fractures in a given set per linear distance measured in a direction normal to the fracture plane. It is a measure of relative frequency of fracturing. A volumetric or areal measure which includes the surface areas of the fractures would be more meaningful, but ordinarily it cannot be made in naturally deformed rocks. It is assumed, therefore, that the relative number of fractures also reflects the relative surface area of these same fractures. The denominator of the fracture number is 100 ft; thus, a fracture number of 100 implies that there is about one fracture per foot in a given set.

Factors Known to Contribute to Fracture Development

Though not restricted to the brittle domain, rupture is everywhere involved in the deformation of brittle rocks. It follows that those parameters that tend to increase rock ductility tend also to decrease the dependency on rupture. Figure 1 is a summary of the deformation characteristics of rock as measured and observed in the laboratory. Each of the drawings represents a generalized final configuration in longitudinal cross section of an originally intact right cylinder of rock material that has been compressed parallel with the cylinder axis. Ductility increases from left to right, as can be seen from the typical stress-strain curves at the bottom of the illustration or from the average permanent longitudinal strain indicated at the top of each of the columns. Very brittle behavior is characterized by a nearly linear stress-strain response up to the rupture point (stage 1). In most cases only extension fractures are observed. This generalization applies only to macroscopic fractures in standard triaxial tests. The sequence of formation between shear and extension fracturing in general is unresolved. If the stress-strain curve departs from linearity prior to rupture, the observable deformation may be caused by shear fracturing that forms a wedge and terminates in an extension fracture (stage 2). The small wedges give way to failure along discrete throughgoing shear fractures (stage 3). At this stage, some observable permanent distortion of the cylinder wall can be seen, and the specimen is said to have "barreled." Under even more ductile conditions (stage 4), the major offset is along a shear zone rather than along a discrete fracture. This zone contains numerous small fractures, all of which can be of the shear type or mixtures of both shear and extension fractures. Very ductile rocks are characterized by large permanent strain without evidence of macroscopic fracture (stage 5).

Extensive laboratory studies of the mechanical properties of common sedimentary rocks have been published (Handin and Hager, 1957; 1958; Handin *et al.*, 1963; Handin, 1966). Though precise constitutive relations are unknown for rock materials because of insufficient data on time effects, these studies have provided an empirical basis for the single effects of several important parameters. The factors that affect rock ductility and, therefore, rupture are: rock type, temperature, effective confining pressure,[3] and strain rate. Thus, there are several ways that the spectrum of behavior shown in Figure 1 can be achieved. Increasing effective confining pressure or temperature or decreasing strain rate tends to increase ductility. Just how much the ductility increases depends on rock type. Quartzites and dolomites, for example, never become as ductile as stage 5 specimens under environmental conditions encountered in sedimentary basins. Limestones, however, may behave like specimens in stages 1 and 2 at or near the surface, but at 25,000 ft (7,620 m) of burial their behavior may be more similar to stage 5 specimens. Contrasted with quartzite, halite is ductile even at atmospheric conditions. Sandstones, depending on the degree of cementation, behave in a fashion intermediate between those of limestones and dolomites. Because of their variable composition and fabric, shales show the widest range of behavior for any given set of conditions. These statements exclude flow by large-scale cataclasis, which can, of course, lead to grossly ductile behavior but which is accomplished in part by rupture in constrained beds.

The change in ductility of common sedimentary rock types is plotted in Figure 2 against depth of burial, using normal gradients of overburden pressure, pore pressure, and temperature. On the basis of laboratory work, very little difference in ductility of the common sedimentary rock types at near-surface conditions would be expected, but, under several thousand feet of burial, large differences should exist.

In natural rock deformation, overburden, pore pressure, and temperature are not independent variables. In order to compare natural rock deformation under the same environmental conditions, one can restrict his observations to narrow stratigraphic intervals and similar structural positions where the effective overburden pressure and temperature would be about

[3] Effective confining pressure is the external hydrostatic confining pressure (analogous to total overburden pressure) less the internal hydrostatic pore pressure (analogous to formation pressure).

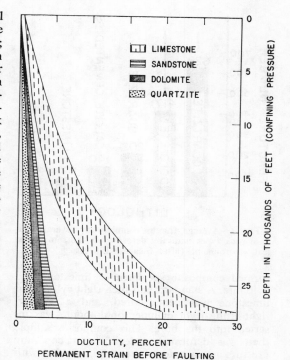

FIG. 2—Ductilities of water-saturated rocks as a function of depth. Effects of confining (overburden) pressure, temperature, and normal formation (pore) pressure are included (after Handin, *et al.*, 1963).

the same. The lithologic effect then can be isolated and studied as a single variable. The results of such a study are shown in Figure 3. The fracture number is computed for many stations in each of the lithologic types and is averaged. These stations are at the same structural position, and they cover a stratigraphic interval that is very small compared to the total depth of burial at the time of deformation. The vertical arrows above quartzite and dolomite indicate that at some stations the rock is too shattered for meaningful measurements. The number above each column indicates the average for that lithologic type exclusive of stations where there is shattering. From these data it can be concluded that, though absolute ductilities (Fig. 2) may differ in nature, primarily because of lower strain rates, the *relative* ductilities of the common sedimentary rock types as measured in the laboratory are validly applicable to natural deformations.

These data become very significant in prospecting for fractured reservoirs. Figure 4A is a thin section made from an experimentally de-

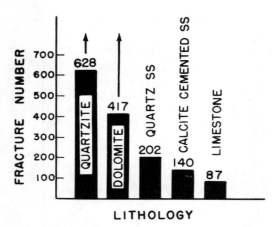

FIG. 3—Average fracture number for several common rock types naturally deformed in the same physical environment (after Stearns, 1967).

formed composite specimen of limestone and dolomite. A standard half-inch right cylinder of limestone was center-bored, and a smaller, tight-fitting right cylinder of dolomite was inserted into the bore. This composite sample then was deformed under 5 kbar confining pressure in an extension test (σ_3 parallel with cylinder axis). The contrasting behavior between the limestone and dolomite is evident in Figure 4A. The limestone deformed by intra-

granular flow mechanisms whereas the dolomite deformed by fracture. Figure 4B shows the same behavioral relation between limestone and dolomite except that they were deformed in a natural environment. It is noteworthy that the vast majority of fractured reservoirs in the Permian basin are in Ellenburger dolomites, not the limestones. From the foregoing data, it is evident that knowledge of the mechanical properties of all common sedimentary rocks, not just limestone and dolomite, should be used actively in prospecting for or developing fractured reservoirs.

NATURAL FRACTURE SYSTEMS
General

Fracture systems that are both pervasive and consistently oriented throughout a large volume of rock can be subdivided into two major classes: regional orthogonal fractures and structure-related fractures. Without further explanation this categorization can be ambiguous. Vast areas commonly may contain one or more fracture patterns, each of which is composed of two very regular and continuous fracture sets that are normal to one another. Their restored orientations are independent of local structure, and these fractures are commonly well developed, even in flat-lying beds. Because all fracturing results from a stress difference, the ori-

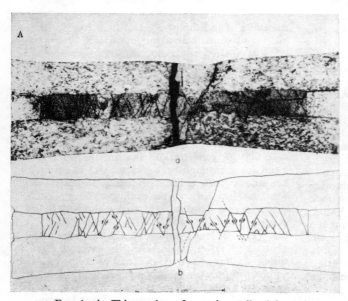

FIG. 4—**A.** Thin section of experimentally deformed composite specimen of limestone (upper and lower units) and dolomite (middle unit; after Griggs and Handin, 1960). **B.** Sequence of naturally deformed limestone (upper and lower units) and dolomite (middle unit). Rock type in either **A** or **B** can be recognized on basis of fracture number.

gin of these fractures must be regarded as a problem of structural geology. However, these fractures are excluded from the structure-related class because they are not the result of stress states associated with specific geologic structures. Rather, they are somehow involved in the structural development of an entire region and are classified, therefore, as regional fractures. Structure-related fractures, however, are associated with specific features like folds or faults. Contributing to the ambiguity of this classification is the fact that these structure-related patterns can pervade entire regions where similar parallel structural features also are common. However, any local change in the structural trend changes the trend of the restored orientation of structure-related fractures, whereas restored regional fractures would be unaffected. The structure-related fracturing is a deformation mechanism within a specific structure; it can be homogeneous only where the local structures are also homogeneous throughout the region.

Regional Orthogonal Patterns

Many areas like the Colorado Plateau, Uinta basin, and Piceance Creek basin are characterized by one or more sets of regional orthogonal fractures. In the Uinta basin, for example, a ubiquitous set of fractures strikes nearly north-south. A second set strikes east-west. Both sets are found throughout the basin, are independent of local structures, and are thus regarded as regional orthogonal fractures. They are rotated in places by later folding but, where the beds return to an undisturbed position, the fractures return to their monotonous regional orientations. Throughout the basin, the intersection of the two sets remains normal to the bedding, and any change in strike is due to bed rotation by later folding.

Speculative origins of regional orthogonal fractures vary; they have been discussed by several writers (Blanchet, 1957; Price, 1959; Hodgson, 1961). Low-level cyclic stress differences like those due to lunar tides, which could cause fatigue failure in rocks, are appealing (Blanchet, 1957), but unlikely, sources for regional orthogonal fractures. In some areas, regional fractures are not developed at all, even though the lithologic types are similar to those of other localities where such fractures are found. The lunar tides, of course, affect the entire surface of the earth. Furthermore, two orthogonal patterns (four fractures) which are commonly found in the same area cut rocks of

Tertiary age. If lunar tides were the cause, the existence of two patterns would imply a drastic change in the tidal direction.

Hodgson (1961) proposed that regional fractures are inherited, *i.e.*, the pattern is propagated upward into younger beds as they become lithified. This explanation is made doubtful by field evidence from at least one area. On the Uncompahgre Plateau, a regional orthogonal fracture pattern is present in the lower and upper beds of the Jurassic System but is absent in the intervening layers. Here the Entrada Sandstone is a calcite-cemented sandstone about 125 ft (40 m) thick. It is completely free of fractures. Indeed, local inhabitants refer to it as "slick rock." The underlying Kayenta and the overlying Summerville sandstones both contain the same regional orthogonal fracture patterns. That the fractures in the Summerville could be inherited from the underlying beds through a completely unfractured layer 125 ft thick seems most implausible.

Price (1959) gave theoretically, but not always geologically, sound arguments for the formation of regional orthogonal fractures. His mathematical arguments for an origin by uplift are logical and fit certain geologic facts—namely, the common association with plateau-type uplift. Price's arguments, however, do not explain the occurrence of two superposed patterns (four fractures).

Even though none of the explanations for regional orthogonal fractures is satisfactory, several empirical statements can be made concerning their occurrence. These fractures are commonly continuous as a single break or a narrow zone for long distances. On the Colorado Plateau, for example, patterns that are continuous for several miles can be seen on air photographs. If such a fracture were intersected by a borehole, it could greatly affect fluid communication to the well.

Vertical continuity depends on the nature of the layered sequence. These fractures rarely terminate vertically within a bed, but continue vertically through several beds. They generally, but not everywhere, terminate at the contact with a shale unit. In many areas the fractures transect thick units of shale and sandstone with no refraction at the lithologic interfaces. Outcrop conditions usually limit measurements of total vertical continuity, but fractures that cut 400–500 ft (120–150 m) of section are not uncommon. Thus, holes deviated through a producing interval to intersect several regional orthogonal fractures should improve well performance. Furthermore, the orientations of regional orthogonal fracture systems are extremely consistent, and, where they are known from core or outcrop studies, the best direction for deviation can be established.

Fractures Associated with Faults

Fractures associated with faults are generally assignable to the same stress state that caused the fault. The shear fractures are miniatures of the fault. It is not surprising, therefore, that their orientations, as well as that of the associated extension fracture, can be predicted from knowledge of the attitude of the fault. The converse is true also: knowledge of the fracture orientations reveals the strike of the fault as well as its sense of shear.

The orientations of fractures strictly associated with faults change with the fault attitude. This statement holds for both dip and strike. What cannot be predicted for any fault is the relative development of the three potential fractures or the width of the fractured zone parallel with the fault plane. There is no way to estimate the number of fractures that will be associated with a given fault. The total displacement on a large fracture can be accomplished by cumulative slip on several smaller parallel fractures. The amount of throw on a fault lends no insight either. Some large faults have narrow zones of fracturing whereas many small faults have wide zones of small-scale rupture. Furthermore, the shear fracture parallel with the fault is not everywhere well developed; the conjugate shear fracture is commonly more conspicuous.

However, the most likely direction of fluid communication that results from fractures associated with faults can be determined. The strikes of all three potential fractures, as well as their intersection, parallel the faults, so the direction of lateral communication can be predicted. The proper deflection of a hole to intersect the greatest number of fractures depends on the attitude of the fault and the relative development of the associated fractures. For low-dipping faults (of the order of 30°), no deflection would theoretically increase the likelihood of intersection. For vertical faults, deflection either away from or toward the fault in a direction normal to its strike would increase chances of intersection. For normal faults, knowledge of the relative development of the two shear fractures is required. If the fracture that parallels the fault is better developed, deflection toward the fault in the downthrown block and

away from the fault in the upthrown block is most likely. If the fracture conjugate to the fault is better developed, drilling in the reverse directions in the two blocks has the best chance of intersecting the greater number of fractures. Core or outcrop analyses are the best methods upon which to base a decision.

Unlike shear fractures associated with folding, many that are associated with faulting display small visible offset. Movements along the shear fractures can act either to increase or decrease the porosity and permeability. If the movement causes rigid-body rotation of discrete blocks along the fractures, permeability and porosity in the fault zone are increased. Because fracture surfaces are rarely perfectly planar, the displacement can cause a poorer "fit" between adjacent sides of the fracture and thus increase fluid flow along the fracture. If there is also fragmentation into chip-sized material along the fracture, these chips can prop any openings. Decrease of fluid flow, both across and along the fractures, may result from mylonitization, crushing, or smearing—especially if the reservoir is a dirty sandstone. One example is the North dome of Kettleman Hills, where shear fractures are associated with normal faults. In the Etchegoin Formation, shear fractures show very little offset, yet they contain zones of accumulated clay and clay-sized material. This fracture-zone material is so resistant to weathering that the fracture zones stand out in relief relative to the unfractured rock. These fractures would most probably be permeability barriers, because the displacements are less than the formation thickness. If an area is crossed by more than one fault trend, a series of isolated production blocks can result. In these more complicated cases of multiple faulting, the procedure of fracture analysis is essentially the same as for a single trend; nevertheless, additional effort should be made to determine not only the several ages of faulting but also ages relative to oil migration. The nature of the fracture zone associated with any particular fault is best determined from outcrop studies, but considerable information also can be gathered from cores (Friedman, 1969).

A second complication arises where folding is associated with the faulting. All the fracture patterns found on folds then have to be considered along with those associated strictly with the fault. If the only folding is slight drag along the fault, it may have little influence on the pattern. However, there may be a rotation problem. If the fractures associated with the fault-

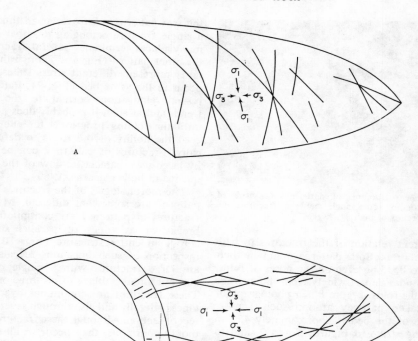

FIG. 5—Schematic illustration of most common fractures associated with a fold. **A.** Pattern 1. **B.** Pattern 2. Both patterns maintain consistent relation to bedding, but not to fold trend.

ing formed early, before much drag had occurred, they would have to be rotated later.

The mechanical interpretations of fractures associated with faults are not a subject of this paper. However, fracturing is best understood in light of the mechanical properties of the rocks (Handin and Hager, 1957), the mechanics of faulting (Anderson, 1951; Hubbert, 1951; Hafner, 1951; Sanford, 1959), and field studies of fractures (Stearns, 1967).

Fractures Associated with Folds

Fractures associated with faults are related to the same stress state that caused the fault. This relation is not true of fractures associated with folds, because analyses indicate that any particular volume of rock can undergo several different stress states through the folding history.

Five major fracture patterns have been found to be associated with folds (Stearns, 1964; 1967). Though all the patterns are significant to an understanding of the folding process, only two have fracture numbers high enough to warrant consideration here. The first pattern consists of two conjugate shear fractures and an extension fracture which indicate that the intermediate principal stress axis (σ_2) is normal to bedding, the greatest principal stress and least principal stress axes (σ_1, σ_3) are in the plane of bedding, and σ_1 is the dip direction (Figs. 5A, 6). On the flank of the anticline the extension fractures of this pattern are in the *ac* geometric plane. This pattern represents shortening in the dip direction, elongation in the strike direction, and no change normal to bedding. The second pattern also consists of two conjugate shear fractures and an extension fracture, thus indicating that σ_2 is still normal to bedding and σ_1 and σ_3 are still in the bedding plane—but σ_1 is parallel with strike and σ_3 is the dip direction (Figs. 5B, 7). On the anticlinal flank the extension fracture of this pattern parallels the *bc* geometric plane. The geometry of pattern 2 represents elongation parallel with dip, shortening parallel with strike, and no change normal to bedding.

That these fractures are the result of the folding process itself and not of the regional stresses that initiated folding is best seen from

FIG. 6—Shear fractures of pattern 1 on flank of fold. Bisector of acute angle between fractures defines dip direction of bed.

the consistent relation of the fractures to bedding orientation. Both patterns maintain their relations to bedding even on the noses of folds, where attitudes depart widely from the average trends of the folds. Figure 8 is a photograph of bedding taken partway around such a nose. The strike of the bed is approximately 45° to the bedding strike in Figures 6 and 7, and to the anticlinal axial trend. The fractures are the shears of pattern 1. Notice that the bisector of the acute angle between the shear fractures remains in the dip direction, and the intersection remains normal to bedding despite the difference between the strike of the bed and the fold axis. A consistent relation is equally true for pattern 2. Therefore, all six fractures are related to the local rotation axis, not to the average trend of the fold axis. The same conclusion is supported by studies of folds that are sinuous along strike. A slight change in strike marks a slight change in the geographic orientation of the fractures, but not in their relation to bedding.

The shear fractures of these two patterns rarely show visible offset. They are designated as a shear or extension type solely on the basis of the geometry of the assemblage, but the consistent and ubiquitous evidence, from both the laboratory (Handin and Hager, 1957; Handin, 1966) and the field (Sheldon, 1912; Melton, 1929; Cloos, 1936; Friedman, 1964; Stearns, 1964; 1967; 1969), cannot be denied. The relative ages of the two patterns cannot be determined absolutely, but pattern 1 probably begins to form earlier than pattern 2. This relation is based on the observation that pattern 1 commonly is present on low-dipping folds without much cross-sectional curvature, even though pattern 2 is absent. However, development of

the two patterns must overlap in time, because a single fracture belonging to either pattern in many places terminates several fractures from the other pattern (Fig. 9). Though the two patterns represent different stress states, they can occur in the same bed (Fig. 9). Pattern 2 represents elongation normal to the anticlinal trend, and as such it probably does not develop until the folding progresses far enough to exceed the ability of the rocks to deform elastically. Fractures of pattern 2 can be explained by large-scale cataclastic flow of the folded sequence in bulk (Stearns, 1969).

The morphologies of the fractures of the two patterns are somewhat different. Many of the fractures of pattern 1 are continuous as single breaks or as zones of parallel subfractures across an entire structure (Fig. 10). Ground inspection of such lineations reveals that they are large fractures with a single orientation, and not an assemblage of all three orientations. These fractures are continuous over large distances laterally and also through several hundred feet of section. Although these fractures are commonly enormous, they occur in all sizes down to the scale of ruptured sand grains. Their homogeneous orientation on all scales is remarkable. Fabric diagrams of pattern 1 shear fractures look the same whether they represent fractures seen in quartz grains or lineations from air photographs. The larger of these fractures could affect fluid communication over vast distances.

The fractures of pattern 2 never attain the very large sizes of pattern 1. They range up to several tens of feet long, but most are only a

FIG. 7—Fractures of pattern 2 on flank of fold. Horizontal line (W15) is parallel with bedding strike and is 18 in. long. Bisector of acute angle between shear fractures parallels bedding strike.

FIG. 8—Shear fractures of pattern 1 on nose of anticline (position of photograph relative to overall fold is indicated on sketch). Pen parallels bedding strike, which is at angle of about 45° to anticlinal trend.

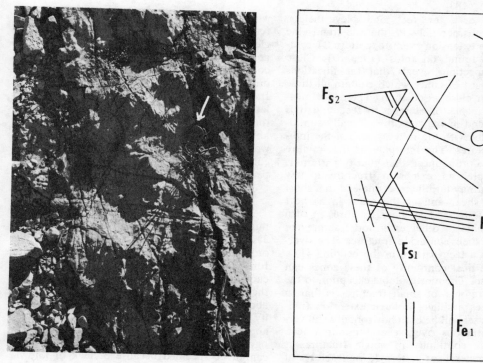

FIG. 9—Photograph and sketch of fractures of both patterns 1 and 2 in same bed. Shear fractures, pattern 1 (F_{s1}); extension fractures, pattern 1 (F_{e1}); shear fractures, pattern 2 (F_{s2}); extension fractures, pattern 2 (F_{e2}).

FIG. 10—View of anticline showing shear-fracture zones of pattern 1 (dark lineations form acute angle in dip direction) and zones of pattern 2 (lineations of trees are parallel with anticlinal trend). Pattern 2 zones are made up of three fracture sets—two shear fractures and an extension fracture.

few inches or a few feet long. Nevertheless, long lineations parallel with the fold trend (the trend of the extension fracture, pattern 2) commonly are found on folds (Fig. 10). Close ground inspection reveals that the lineations usually consist of zones which contain all three fractures in close association. These are measurable in inches or at most in feet, and it is only the trend of the zone that is really continuous parallel with the fold strike, not the individual fractures. The interplay of the fracture sets is more pronounced than that in pattern 1. For example, an extension fracture a few inches long may terminate at one end in a short left-lateral shear and at the other in a short right-lateral shear. The shear fractures in turn may terminate in other extension fractures or perhaps in their conjugate fractures. The overall effect is a lacy pattern in the bedding (Fig. 7). The vertical continuity of these zones can be large also. The most notable exception to the common presence of small fractures is that in a few places a single, isolated extension fracture of this set will be both horizontally and vertically continuous over several hundred feet. Though the continuity of single fractures is greater in general for pattern 1, the fractures of pattern 2 may be more effective fluid passages. There are no data to support this idea, but it is plausible. Because the fractures of pattern 2 are smaller, local rotations of the small blocks

within the zone are likely. This rotation would result in a more open system than that of a single fracture or of elongate parallel slivers. The dense fractures of pattern 2 also may reflect considerable elongation locally that results in dilation (decrease of bulk density) along the fracture zone. Because of the large size and isolation of the fractures in pattern 1, there are three possible directions of communication in any well that intersects them. However, for wells intersecting pattern 2 fracture zones, communication is favored in one direction— parallel with the structural trend.

Factors that Affect Number of Fractures

All other factors being constant, fracture number is most affected by rock type (Figs. 2, 3). This is true of both patterns 1 and 2, but only if fracturing took place at sufficient depth for the effective confining pressure to be significant to the behavior of rocks. Figure 2 shows that under near-surface conditions all rocks are about equally brittle. The divergence in ductility is the result of effective confining pressure, and only under sufficient overburden is lithology an important factor. Exact depth control cannot be specified because of lack of knowledge of the effect of time. However, the laboratory work of Heard (1962) indicates that reduction of strain rates even seven orders of magnitude has little effect on the brittle-ductile transitions at the relatively low temperatures encountered in oil wells.

A second factor that affects fracture number is bedding thickness. It is known from observation that, for given rock type, structural position, and depth of burial, a thin bed will contain more fractures than a thick bed under the same conditions (Harris *et al.,* 1960). A possible explanation is that fracture spacing depends in part on instabilities developed in the extended multilayered medium. Such instabilities depend on bedding thickness as well as contrasts in physical properties of beds (G. M. Sowers, personal commun., 1969).

Thickness and lithology affect fracture numbers in both patterns 1 and 2. However, structural position is also significant to pattern 2, because here fracturing is an active mechanism of cataclastic flow. The intensity of fracturing is greatest where the rate of change of dip is greatest. This parameter should not be confused with the steepness of dip. Steep dips can result simply from rigid-body rotations between hinge points, and strain need not be large. However, changes in dip over short lateral distances can result in large local strains (elongation or shortening). The elongation can be accomplished by fracturing (pattern 2) along zones parallel with the fold hinge. In these zones, fracture numbers as high as 15,000 have been recorded (Stearns, 1969). In some hinges where the dip change is very great over short distances (60° or more in a few hundred feet laterally), the rock units, including limestones, can be completely shattered by unordered fracturing. The lack of order may result from numerous local stress concentrations developed at fracture intersections.

INFLUENCE OF FRACTURE ON RESERVOIR POROSITY AND PERMEABILITY

Estimate of Fracture Porosity

Fracture width, area, spacing, surface roughness, and filling are the primary factors governing fracture porosity. Reported values of fracture porosity are low, from <0.05 percent (Snow, 1968) to 6 percent (Regan and Hughes, 1949, p. 47). These low values are not surprising if one realizes that a porosity of less than 2 percent is provided by unfilled fractures 0.04 in. (1 mm) wide developed around the six faces of 1 cu ft of rock. The ability to estimate fracture porosity is, however, essential to reservoir evaluation. Several approaches to this problem have been reported, and a few examples are discussed below.

Elkins (1953, p. 181–182) measured fracture widths in cores of Spraberry Sandstone and estimated the spacing of vertical fractures from their observed frequency in 3.5-in.-diameter (8.9 cm) cores from five wells and from pressure-buildup data. He found fracture widths up to 0.013 in. (0.33 mm), but the average was 0.002 in. (0.051 mm); he concluded that spacing ranged from a few inches up to a few feet. From these data he calculated a fracture volume of 110 bbl per acre-foot.

A geometric approach applicable to folded rocks was developed by Murray (1968, p. 60). On the assumption that extension fractures form in the outer layers of curved beds, he showed that the fracture porosity varies directly as the product of bed thickness and curvature, *i.e.,*

$$\phi f = \frac{T}{2R} = 1/2T\,\frac{d^2z}{dx^2}, \qquad (1)$$

where T is thickness of bed, R is the radius of curvature, and d^2z/dx^2 is the curvature (dz/dx is dip). This approach is particularly attractive because the calculation is independent of fracture spacing, although spacing is considered in deriving the equation. Moreover, R can be determined from dip information because it is the reciprocal of d^2z/dx^2. Murray applied this analysis to the Sanish pool, McKenzie County, North Dakota, and demonstrated a good coincidence between areas of maximum curvature and areas of best productivity. Using maximum and minimum values of T and R, respectively, he calculated a fracture porosity of 0.05 percent.

Snow (1968) described a method for calculating fracture porosity, openings, and spacings from permeabilities measured in drillholes. The method applies only to rocks with negligible intergranular porosity; it is not valid if solution enlargements, porous interbeds, breccia zones, or fracture spacing less than about a third of the separation between straddle packers is present. Results from studies of selected dam sites indicate maximum near-surface fracture porosities of 0.05 percent. These decrease by an order of magnitude each 200 ft (60 m) within the depth of usual dam-site exploration. Fracture widths are typically 0.004 in. (0.101 mm) at 50 ft (15 m) to 0.002 in. (0.051 mm) at 200 ft (60 m). Average fracture spacing is 4 ft (1 m) near the surface and increases to 14 ft (4 m) at depths of 300 ft (90 m).

Fracture porosity also can be estimated from conventional electric logs where certain conditions obtain and the porosity of the intact reservoir rock is known (Pirson, 1967; Pirson et al., 1967). A fracture-intensity index, FII, is related to fracture porosity through total porosity, ϕt, and the porosity of the unfractured reservoir rock, ϕb, as follows:

$$FII = \frac{\phi l - \phi b}{1 - \phi b} . \qquad (2)$$

Pirson et al. (1967, p. 6) demonstrated that FII is calculated from relations between the resistivities of the mud filtrate, the fracture system filled with mud filtrate, the intact rock, the formation water, the invaded zone (from the short-normal curve), and the zone of 100 percent water saturation (from the long-normal or induction curves). The fracture porosity then can be determined if ϕb is known. Applying this method to a section of Austin Chalk from a producing well in the Salt Flat–Tenney Creek field, Caldwell County, Texas, they calculated a fracture porosity of 0.2 percent.

Estimates of Fracture Permeability

Although the absolute porosity provided by fractures is low, the effective porosity is high because the available fracture-void volume is connected. It follows that the influence of fracturing on reservoir permeability is most important. Pertinent analyses of fluid flow through parallel plates (or fractures) were given by Lamb (1932), Muskat (1937), Huitt (1956), and Parsons (1966), among others. The equation for volumetric flow rate between two smooth parallel plates (after Lamb, 1932) is the common point of departure as follows:

$$q = \frac{-g_c b W_f^3}{12\mu} \frac{dp}{dL}, \qquad (3)$$

where:

q = volume of flow per unit length
W_f = width of fracture
b = depth of fracture
dp/dL = pressure gradient in direction of flow
g_c = a conversion factor
μ = absolute viscosity of fluid.

This equation is approximately valid for viscous, laminar flow for variable values of W_f, provided that the gradient dW_f/dL is small; it applies even if both bounding surfaces are curved (Huitt, 1956, p. 259). Huitt considered the influences of surface roughness and turbulent flow and concluded that, if flow in fractures is laminar, surface roughness has no appreciable effect on the resistance to flow. Surface roughness becomes a significant factor where the flow is turbulent. From consideration of production rates from oil wells, however, both Huitt and Parsons regarded turbulent flow in fractured reservoirs as unlikely.

Baker (1955, p. 384) illustrated the importance to permeability of fracture width, using the basic relations of equation (3). For example, a single fracture 0.01 in. (0.25 mm) wide has the equivalent permeability of 454 ft (188 m) of unfractured rock with a uniform permeability of 10 md; an 0.05-in-wide (1.27 mm) fracture is equivalent to 568 ft (173 m) of rock with a permeability of 1,000 md.

Parsons (1968) modified equation (3) so as to include the permeability of the total fracture-intact rock system in which vertical fractures occur in sets of specified spacing and orientations relative to the overall pressure gradient. He expressed the total permeability of the fracture-rock system as:

$$k_{fr} = k_r + \frac{Wa^3 \cos^2 \alpha}{12A} + \frac{Wb^3 \cos^2 \beta}{12B} + \cdots, \qquad (4)$$

where

k_{fr} is the permeability of the total system
k_r is the permeability of the intact rock
Wa, Wb are the widths of fractures in sets a, b, etc.
α, β are angles between fracture sets a, b, respectively, and overall pressure gradient
A, B are the fracture spacing (perpendicular distance between fractures for sets a, b, etc.).

Further insight into the role of fractures in reservoir permeability has been gained through study of electrical analog models. McGuire and Sikora (1960) showed that the width of artificial fractures is much more important than their length in affecting communication between natural fractures. However, little is gained by widening isolated short fractures because the unfractured formation beyond their ends controls the rate of flow. Huskey and Crawford (1967) kept their simulated fractures at constant width but varied fracture shape, orientation, density, and total length. They found that fracture density correlated closely with production capacity and that the influence of fracture shape was small. Fractures oriented parallel with the isopotential lines [$\alpha = 90°$, equation (4)] do not increase the effective permeability.

In summary, the existing knowledge of fluid flow in fractures provides at least a first approximation of measured flow rates and, what is important here, an appraisal of those attributes of fractures that are truly significant to the problem. The permeability is directly proportional to the cube of the fracture width, inversely proportional to the spacing between fractures, and dependent upon the fracture orientation relative to the direction of the pressure gradient. The permeability decreases with fracture filling and roughness of the fracture surfaces (where the flow is turbulent). It follows that the permeability of a fracture system can be expected to be greatest where the reservoir bed contains wide, closely spaced, smooth fractures oriented parallel with the fluid pressure gradient.

Estimates of fracture permeability range from a few millidarcys to many darcys. Methods used to estimate fracture permeability include pressure-buildup studies, interpretations of volumes of oil produced relative to production from unfractured reservoirs of known permeability, and certain geometric relations. Elkins (1953, p. 184–186) used pressure-buildup data to calculate the average effective permeability in the Spraberry Sandstone reservoir (16 md). This permeability is the same order of magnitude as that deduced by Dyes and Johnston (1953) from a similar analysis of the Spraberry trend. Elkins tested his result by calculating the fracture widths and spacings that would yield this permeability. He found that the 16-md permeability would be provided by fractures .0011 in. wide (0.28 mm), spaced 4 in. (10 cm) apart, or by .0015-in.-wide (0.38 mm) fractures 10 in. (25 cm) apart, or by .002-in.-wide (0.51 mm) fractures 24 in. (61 cm) apart. These predicted widths compare favorably with the average of 0.002 in. actually measured in cores. Some other analyses pertinent to pressure-buildup data are by Miller *et al.* (1950) and Adams *et al.* (1968).

Permeability estimates commonly are made from the volume of oil produced. Regan (1953, p. 213) estimated a maximum permeability of 35 darcys and an average of 10–15 darcys from production rates for the chert zone early in the history of the Santa Maria Valley field, California. He did this by comparing the production rates from the fractured chert with those from "oil sands" in various other fields.

Murray (1968, p. 60–61), reasoning from equation (3), demonstrated that, in folded beds with extension fractures normal to bedding and parallel with the fold axis, the permeability and the porosity (see preceding discussion) are functions of bed thickness and curvature, as follows:

$$K = \frac{A^{\frac{3}{2}}}{48T^2}\left(T\frac{d^2z}{dx^2} \right)^3, \qquad (5)$$

where T is the bed thickness, d^2z/dx^2 is the curvature, and A is the average area per fracture perpendicular to the direction of flow. After conversion to millidarcys and evaluation of A in terms of an assumed fracture spacing of 6 in. (15 cm), equation (5) reduces to:

$$K = 4.9 \times 10^{11}\left(T\frac{d^2z}{dx^2} \right)^3. \qquad (6)$$

The following examples of permeability values were given by Murray (1968, Table 1, p. 61):

$T = 5$ ft		$T = 10$ ft	
$\dfrac{d^2z}{dx^2}(\times 10^{-5})$	K (md)	$\dfrac{d^2z}{dx^2}(\times 10^{-5})$	K (md)
1	0.06	1	0.49
2	0.49	2	3.92
4	3.92	4	31.20
6	13.20	6	106.00

He demonstrated that zones of high productivity in the Sanish pool coincide with those in which the curvature exceeds a certain minimum value. This fact would seem to verify his approach; however, an independent evaluation of permeability was not given.

As stated, fractures can either enhance or restrict fluid flow. McCaleb and Wayhan (1969) reported an interesting case in which both restriction and increase in fluid communication as a result of fractures are found in the same reservoir. At the Elk Basin field, fractures form a pressure-communication barrier in the "A" zone which causes an abrupt pressure drop of 2,000 psi across the fractured area. However, they reported that in other areas the "D" zone has a sufficiently high fluid movement to be assignable as "a fracture-permeability reservoir" (McCaleb and Wayhan, 1969, p. 2111). Emmett *et al.* (1969, p. 3) discussed fractures in the Little Buffalo Basin field (Tensleep reservoir) that are so open that ". . . upon completion the well produced 100-percent water even though analysis of the matrix rock indicates it should be oil productive." McCaleb and Willingham (1967, p. 2126), in their study of the Cottonwood Creek field, stated: ". . . where developed, the fracture systems probably offer the primary means available for fluid movement in the reservoir."

DETECTION OF FRACTURES

Early recognition of a fractured reservoir will influence the location and number of subsequent development wells, and, therefore, is of major economic significance. Some of the methods used to detect and study fractures *in situ* are listed below in the approximate order that they would be used.

1. Loss of circulating fluids during drilling is widely recognized as a positive indication that a fractured or cavernous formation has been penetrated. (Very narrow fractures that can greatly increase the effective permeability may be missed, however.) The axiom is (Daniel, 1954): field wells in which a large volume of mud is lost almost without exception become the major producers if they produce at all.

2. Fractures in cores provide direct information on fracture development if natural fractures can be distinguished from those induced by coring, and if fracture spacing and orientation permit adequate sampling in cores (Friedman, 1969, p. 369–375). For some rocks, poor core recovery suggests intensely fractured zones. At least, porosity and permeability data for unfractured samples of the reservoir rock are obtained from the cores. These measurements are preferably made at the pressures of the reservoir (Fatt and Davis, 1952; Fatt, 1953, 1958).

3. Well-test analyses provide at least two types of data from which the influence of fractures can be evaluated: production rates and pressure-buildup interference. Perhaps the most common clue to a fractured reservoir is a flow rate many times that to be expected from the porosity and permeability data of the unfractured reservoir rock.

Use of pressure-buildup data was reviewed by Odeh (1964), Matthews and Russell (1967), and Adams *et al.* (1968), among others. Pressure-buildup data consist basically of well-bore pressure versus shut-in time (Horner, 1951; Black, 1956; Gray, 1962). The shape and slope of the observed curve or line are compared with those expected from various models of the reservoir. For example, Warren and Root (1963) analyzed such data in order to determine both intergranular porosity and fracture or vugular porosity on the assumption that the former contributes significantly to the pore volumes but negligibly to the flow capacity. Adams *et al.* (1968, p. 1193), using 2-week pressure-buildup tests run every 2 to 5 months for several years, recognized sharp downward bends in the pressure-time curve. They stated that ". . . although many factors—faults, stratification, heterogeneity, fracturing, etc.—are known to cause a pressure buildup curve to

bend upwards, very few effects other than a permeability increase or other improvements in rock flow character are known to cause a decrease in the straight line slope." They interpreted the increase in effective permeability to be caused by a fracture (or other discontinuity) at some distance from the well bore. The time of the buildup-curve bend can be used to estimate the radial distance to the fracture. This approach is one of the few techniques capable of detecting the influence of fractures away from the borehole, and it is widely used.

4. Certain well-logging techniques may be helpful in recognizing and locating fractures (Pickett and Reynolds, 1969). Pirson (1963, 1967), Nechai and Mel'nikov (1963), and Pirson *et al.* (1967) discussed the use of conventional logs to evaluate fracture porosity and thus to detect fractures (see equation 2). Sonic and sonic-amplitude logs have been used to locate fractures, but their interpretations are complicated by many factors (see review by Morris *et al.*, 1964).

5. Several downhole direct and indirect viewing systems, including downhole photographic and television cameras, packer impressions, and "acoustical picture" reconstructions, have been used to detect fractures on the borehole wall. All these devices include a system for obtaining the azimuthal orientation of the record. Direct optical systems employing 16 mm and 35 mm multiframe borehole cameras have been described by Dempsey and Hickey (1958), Jensen and Ray (1965), and Mullins (1966). Briggs (1964) described a television system for immediate viewing of the borehole wall. Each of these visual systems requires a transparent environment and light source. Each is limited in depth by ambient temperatures. Packer-impression techniques require the inflation of an impressionable tubing against the walls of the borehole (Fraser and Pettitt, 1962; Anderson and Stahl, 1967). When the tubing is partly deflated and withdrawn from the borehole, the trace of a fracture can be observed where the tubing had been injected into the opening.

The Borehole Televiewer described by Zemanek *et al.* (1969) uses an entirely new concept to provide a picture of the reservoir rock. A piezoelectric transducer probes the borehole wall with acoustic energy. The amount of energy reflected back to the transducer is a function of the properties of the rock. In general, irregularities reduce the amplitude of the reflected signal. Return signals are observed on an oscilloscope and photographed for permanent record. A flux-gate magnetometer senses the earth's magnetic field and provides the means for determining the orientation of the log. As with the camera techniques, features on the cylindrical borehole are observed only in two dimensions (Fig. 11). A typical Borehole Televiewer log (Fig. 12) clearly shows fractures and small vugs. The tool now can be used at temperatures up to 300°F and in a borehole filled with any homogeneous, gas-free liquid like fresh water, saturated brine, crude oil, or drilling muds. All these techniques give information on fracture development directly adjacent to the borehole. When used in combination with well-test analyses, they can provide the confirmation of fracturing necessary for complete reservoir evaluation.

6. One common characteristic of fractured reservoirs is that the productivity of wells is greatly increased by artificial stimulation (Elkins, 1953; Walters, 1953; Hunter and Young, 1953; Daniel, 1954; Baker, 1955; Hassebroeck and Waters, 1964; and Shearrow, 1968). For example, in the Spraberry Sandstone of West Texas, initial potentials of wells range up to 1,000 bbl/day after fracture treatment as compared to estimated capacities of 5–10 bbl/day if oil had to flow into the borehole through the siltstone itself.

Acidizing of fractured carbonate reservoirs is done primarily to increase the width of natural and artificial fractures and the size of the borehole. Increasing the diameter of the borehole enhances the probability of intersecting natural fractures. In addition, Baker (1955, p. 384–387) showed that for oil with 50 lb–ft³ density and 1 cp viscosity, channeled to a 6-in. (15 cm) diameter borehole through a horizontal fracture 0.04 in. (1.00 mm) wide with a radius of 660 ft (200 m), most of the pressure drop occurs within a foot of the hole. An acid treatment to increase the borehole diameter to 9 in. (23 cm) would reduce the total pressure drop and increase the flow capacity significantly.

7. Other techniques for fracture detection become useful where several wells penetrate the same reservoir. All are based on evidence of fluid communication between wells which can be explained only by the presence of fractures. Pressure interference is perhaps the most common of these, and it generally is cited as clear-cut evidence of a fractured reservoir (Elkins, 1953, 1960; O'Brien, 1953; Baker, 1955). In the Agha Jari field, Iran, pressures in two ob-

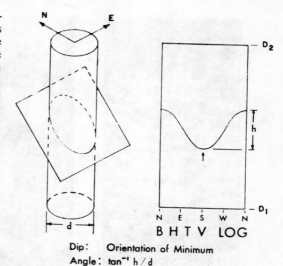

Dip: Orientation of Minimum
Angle: tan⁻¹ h/d

Fig. 11—Isometric sketch of fracture or bedding plane intersecting borehole at moderate dip angle, and corresponding Borehole Televiewer log (after Zemanek et al., 1969, Fig. 7).

servation wells, 11.5 mi (18.4 km) apart, fell at nearly the same rate within a lag of only a few pounds per square inch (Baker, 1955, p. 386). Similar evidence of communication over distances of 40–50 mi (64–80 km) have been reported (O'Brien, 1953; Baker, 1955). Elkins (1960) used pressure-interference data to determine the effective fracture orientation in parts of the Spraberry field. Furthermore, from the observation that all development wells within a given area in the Spraberry field exhibit reduced initial reservoir pressures (i.e., they all drain the same interconnected system of fractures), he concluded that the number of such wells could have been reduced and their spacing increased.

Other methods in this category include monitoring water-breakthrough times in waterflood tests and gas-injection tracer tests (Barfield et al., 1959, and Elkins, 1960). McCaleb and Willingham (1969) described the Permian Phosphoria reservoir, Cottonwood Creek field, Wyoming, in which the true nature of the reservoir (fracture plus intergranular porosity) was discovered only after poor gas- and water-injection performances were noted. The rapid channeling of injected phases stimulated a review of available geologic, engineering, and production data, which resulted in recognition of the dual nature of the reservoir. Data on communication times and directions relative to

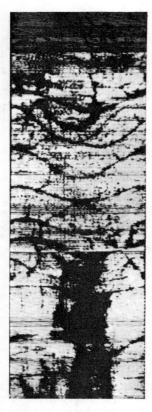

FIG. 12—Borehole Televiewer log in fractured West Texas Precambrian formation. Well bore was filled with drilling mud (after Zemanek *et al.*, 1969, Fig. 9).

surrounding producing wells can be used to calculate effective permeabilities.

TECHNIQUES THAT AID IN SUBSURFACE-FRACTURE INTERPRETATIONS

Fractures in Cores

Studies of cores containing natural fractures can provide statistical information on fracture spacing, width, and orientation. In addition, the effective directional permeability due to small fractures can be measured in places, and indirect methods for fracture detection can be calibrated against the observed occurrence. These studies require, however, that (1) natural fractures are distinguishable from those artificially induced by coring, (2) biases inherent in sampling fractures in cores are considered, and (3) core orientation is known. Care must be exercised from the time the core is permanently removed from the barrel to record whatever information is available on core orientation and to recover and preserve maximum, intact lengths of core. Fractures are best observed if full cores are washed and slabbed along several cuts parallel with the core axis. Very fine fractures can be detected by X-ray techniques or by use of fluorescent-dye penetrants (Gardner and Pincus, 1968).

There commonly is a problem in distinguishing natural fractures from those induced by coring, from partings along bedding planes, or from separations that might have occurred during handling of the cores. A workable classification is as follows (Friedman, 1969, p. 369): *unequivocal* natural fractures are partly or completely filled with gouge or vein material or are unfilled but parallel with nearby filled fractures (including microfractures in individual grains as observed in thin section); *very probable* natural fractures are those with slickensided surfaces and fractures parallel with them; *probable* natural fractures have clean fresh surfaces and are accompanied by parallel incipient fractures (or microfractures) that, in turn, are parallel with unequivocal fractures in the same or adjacent length of core. Criteria to recognize fractures induced by coring are not so straightforward. We have observed cores broken repeatedly by fresh clean fractures that within short distances curve from subparallel to sub-perpendicular to the core axis. These may result from twisting or bending of the rock cylinder during coring. The fractures that produce disked cores are also probably artificially induced. These fractures are evenly and closely spaced and are typically oriented nearly perpendicular to the core axis. The disking may be caused by coring rock in which the *in situ* differential stress is high (Obert and Stephenson, 1965). Disking thus may suggest the possibility of potential casing collapse or formation failure during the production of the well. Though disking may be triggered by the coring, it may occur also along incipient unequivocal natural fractures (Friedman, 1969) or parting planes. We have found that certain phenomena like fracturing, which perfectly parallel the axis of the core or the development of plumose or concentric markings on fracture surfaces, are not reliable criteria for distinguishing between natural and artificial fractures.

Two factors introduce bias in sampling macrofractures (*i.e.*, fractures visible without optical magnification and measurable with hand tools; see below). These are the spacing of the fractures relative to the diameter of the core and the orientation of the fractures relative to the hole deviation (core axis). If, for example, the natural macrofractures are actually vertical and are spaced 1 ft (30 cm) apart, they obviously are poorly sampled by even a 6-in. (15 cm) core bit in a vertical hole. The sampling improves, however, as the angle between the fracture and the core axis increases (Friedman, 1969, p. 374). It is not generally realized that, if a perfectly uniform three-dimensional array of fractures (*i.e.*, fractures developed equally in all possible orientations) is cored, the frequency distribution of observed angles (θ) between the core axis and the normal to the fracture surface (reduced to a two-dimensional plot) is Gaussian, having a peak at $\theta = 45°$. This Gaussian distribution exists because the distribution of angles between fracture surfaces and core axis (after Bloss, 1957) is:

$$P = 100 \int_{\theta_1}^{\theta_2} \sin \theta d\theta = 100 \left(\cos \theta_1 - \cos \theta_2 \right), \quad (7)$$

where P is the probability and $\theta_1 - \theta_2$ is the cell width in degrees; and the preferential sampling of fractures by a core, which increases as the sine of the angle between the core axis and the fracture plane, is:

$$P' = P_\theta \sin (90^2 - \theta), \quad (8)$$

where P' is the final probability and F_θ is the probability at a given θ from equation (7).

Friedman (1969, p. 374) used the relation in equation (8) to weight observed frequencies of different fracture sets in cores from the Saticoy field, California, in order to determine their order of relative abundance. The sampling biases do not influence the study of microfractures from individual sand grains, because the core dimensions are very large compared to the scale of the grains. A sufficient number of microfractures in any possible orientation can be measured in three mutually perpendicular thin sections to give a statistically sound three-dimensional sample. It is also important to realize that poor core recovery from a given interval may imply the presence of a highly fractured zone.

The absolute orientation of fractures in cores can be obtained only when the *in situ* orientation of the core itself is known. First the core must be oriented, then the fractures in the oriented core must be accurately measured. Modern oriented coring devices satisfy the first requirement, although most cores are not oriented during routine operations. Cores not oriented *in situ* can be oriented later from the traces of bedding on the surfaces and from dipmeter and hole-deviation data. Cores taken from holes deviated in excess of 10° from the vertical in thin-bedded, horizontal rocks can be oriented from the elliptical trace of bedding on the deviated core. Even where oriented cores are available, it is a cumbersome task to hold the core in its *in situ* orientation during measurement of the orientations of fractures. Moreover, the fractures are visible only at the surface of the opaque core. In order to alleviate these difficulties, we designed a fracture-orienting device which permits both the orientation in space of a core and the accurate measurement of the strike and dip of fractures or other planar or linear features that transect the core. The method is to (1) insert the core into a suitably sized plastic cylinder, (2) draw onto the surface of the cylinder the traces of bedding and usable fracture planes (Fig. 13A, B), (3) orient the cylinder (now representative of the core) as it was *in situ* from knowledge of hole deviation and strike and dip of bedding (Fig. 14), and (4) measure the attitudes of the fracture traces. A plastic cylinder is used rather than the core itself, because the tracings of planar features clearly define easily measured planes when viewed through the transparent cylinder, and because of the mechanical difficulties involved in supporting and rotating the heavy, oddly shaped core. Cores taken from a

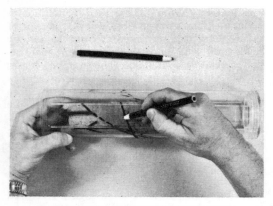

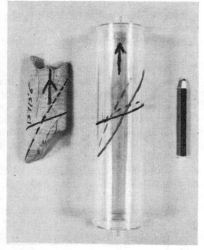

Fig. 13—Photographs of core and plastic-cylinder facsimile: *upper*—Core is held in plastic cylinder of similar diameter while traces of bedding and fracture are marked on surface of cylinder; *lower*—Traces of planar features are clearly visible through plastic cylinder.

deviated well represent the most complicated case for which the apparatus was designed. The apparatus can be used also for cores taken from vertical wells or for cores which previously have been oriented.

The fracture-orienting device was used by Friedman (1969) in the analysis of fractures in cores from the Saticoy field, Ventura County, California. The study shows that both the macro- and microfracture patterns are reasonably homogeneous along the length of the field and that these fracture patterns, as well as small changes in the relative abundance of microfractures, can be used to predict the orientations and relative positions of the two major

faults in the field. The producing sandstones are the lower Santa Barbara and the upper Pico (upper Pliocene); they are not regarded as fractured reservoirs, but both macro- and microfractures do influence productivity. For example, the fractures introduce a marked directional permeability. The permeability across the array of *unequivocal* natural fractures (corehole 2, Fig. 15) is 10 md, that parallel with the fractures and containing the gouge material (corehole 3, Fig. 15) is 40 md, and that parallel with the fractures and not in the gouge (corehole 1, Fig. 15) is 140 md. This differential permeability probably exists in much of both reservoirs because there is a statistically homogeneous fracture pattern developed along the length of the Saticoy field. Emmett *et al.* (1969) also noted a directional effect by fractures on permeability. They reported a 10–30 percent reduction in permeability normal to fractures as compared to permeability parallel with the fractures in the Little Buffalo Basin field.

Fracture Prediction as an Exploration Method

In the introduction we noted that most fractured reservoirs are "discovered" after the initial discovery. We know of no clear-cut example in the published literature of methods of exploration for fractured reservoirs. Therefore, this section contains suggestions as to what could be done rather than methods which have been proved.

Sufficient laboratory data are available to allow construction of fracture-facies maps. The first requisite of such a construction is the ability to establish the burial history at the time of hydrocarbon migration. From burial data, limits can be placed on the effective confining pressure and temperature throughout a basin. Knowing the conditions under which the fracturing had to take place enables one to establish the lithologic types most likely to contain a high density of fractures. This approach can be augmented profitably if precise outcrop studies have been made in the exploration area. Environmental parameters then can be integrated even more closely into the mapping, and absolute rather than relative fracture numbers can be determined for certain rock types or lithofacies. The effect of bed thickness likewise can be evaluated, and those assemblages of rocks that would best serve as fractured reservoirs can be mapped just as any other facies element. Furthermore, if the fractures are structure-related,

Fig. 14—Photograph of fracture-orienting device showing plastic cylinder oriented as corresponding core *in situ*. Borehole deviation is 24° to the S22°W, read at indexes (A) and (B), respectively. Strike of planar features determined by sighting along "strike bar" (C) and read at index (D). Dip of planar feature read at (E).

certain overlay maps may outline targets more exactly. For example, areas of anticipated exceptionally high fracture density, like trends of high rate of change of dip, can be plotted on the fracture-facies map. Overlay maps that include the orientations of various types of fractures (pattern 1 or 2, or fractures associated with faults) may pinpoint specific locations within the potential fracture facies where intersecting fracture zones would occur. Updip limits of fracture facies can be mapped with the same accuracy as for any other facies mapping.

An obvious example of the fracture-facies approach would involve mapping the distribution of dolomitic or siliceous facies relative to limestone or shale facies. If the brittle-ductile facies contact crosses a structure so that the ductile facies is updip from the brittle facies, dry holes in the structurally high areas do not preclude potential discoveries in the downdip brittle facies.

Specific testing of physical properties of lithologic types in a given area also may aid in exploration. If, early in the exploitation of an oil province, a given lithofacies is recognized as a fractured reservoir, then the question arises

as to what other facies may also contain fractured reservoirs. The Sooner Trend on the northeast flank of the Anadarko basin in Oklahoma is a case in point (Durham, 1962; Ward, 1965). Brittle, cherty members of the thick Mississippian carbonate section yield high initial potentials. The cherty members grade laterally into siltstone and shelf carbonate facies.

One approach to the development of such an area would be to restrict drilling to the proved fractured, cherty facies. This strategy would require, however, an accurate knowledge of the distribution of that facies, and other fractured reservoirs might be missed. A second approach would be to investigate experimentally whether other associated facies could be potentially fractured before attempting any stepout drilling. Assume, for example, that facies A forms a fractured reservoir and we wish to determine if facies B and C might also be fractured. Assume also that all three facies occur in the same structural framework and have the same deformation history. Typical specimens of the three facies could be fractured experimentally under properly simulated physical conditions reconstructed from the geologic setting. These

FIG. 15—Photograph of core from Saticoy No. 14, 9,400–9,420 ft, Saticoy field, Ventura County, California, showing location of perm-plugs (1–3) discussed in text. Permeability data were furnished by Shell Development Co.

tests could be repeated at the same temperatures and effective pressures but at different strain rates. If all three facies should prove to be brittle and strain-rate insensitive, but should fracture at different magnitudes of differential stress, certain conclusions could be drawn. If facies A has a higher fracture strength than do facies B and C, it can be concluded that B and C are potential fracture reservoirs. No such prediction could be made for any facies stronger than A. Should a facies fail by ductile flow at a yield stress lower than the fracture strength of facies A, the reservoir quality of that lithofacies would depend on the survival of intergranular porosity during the flow process.

Surface Measurements of Fractures

The purpose of measuring fractures in the field is to establish the orientations and morphologic characteristics of the fracture sets in a given region. In the absence of core data, these surface measurements should be projected into the subsurface. The extrapolation may well be uncertain, but even a qualitative prediction of subsurface conditions is better than none at all.

Obviously, every single fracture cannot be measured or even seen in the field. A major difficulty, therefore, as with fractures in cores, is one of sampling. If the data are to be projected eventually into the subsurface, the collection of these data should be aimed at a good statistical sampling of the following: (1) separation of all fractures into sets and determination of the orientations of these sets; (2) measurements of fracture numbers for each set including variations with bed thickness, lithology, and structural position; (3) establishment of the horizontal and vertical continuities of all sets; and (4) observations of morphologic differences between the sets.

Erosion is a large deterrent to sampling. Highly developed fracture sets or systems commonly accelerate both mechanical and chemical weathering processes, which may result in removal of the fractured rock or in burial under erosional debris. Weathering along fracture zones frequently develops erosional gullies. Debris collects at the bottoms of depressions and obscures the fractures from view. If an area contains parallel gullies, their bottoms should be examined carefully. Of course, the mere existence of parallel drainage systems does not necessarily prove a fracture-controlled origin.

Erosional control by fracture creates still another difficulty of good sampling, namely, that of outcrop boundary orientations. Rock ledges usually spall parallel with major fracture systems. Eventually, ridges develop parallel with these fractures so that the maximum lateral distance of outcrop exposes the least number of fractures. In order to compare the number of fractures between fracture sets, it must be determined that the outcrop examined did not favor the exposure of one particular fracture set. If the outcrop pattern is linear, it is advisable to record its direction. The use of air photographs simplifies this problem. If the preferred development of some particular orientation is suspected but there are too few measurements to be certain, outcrops that favor the suspected fracture set should be sought.

In outcrop studies, the following data should be collected at each station: (1) attitude of bedding, (2) average attitude and variation in attitude of each fracture set, (3) fracture number for each fracture set, (4) lithology, (5) complete description of fracture morphology by set, and (6) an oriented sample (necessary for correlation of the microfracture and macrofracture fabrics).

Fracture number can be recorded as either absolute or relative. An individual worker can develop his own subjective system of relative fracture number such as high, medium, or low, based upon his own impression at each outcrop; arbitrary but consistent limits should be set on the relative values. With experience, this system becomes reliable, but the relative measurements of several workers should not be intermixed. The easiest way to make absolute measures of fracture number is to mark an outcrop with a tape in any convenient direction, independent of the orientation of the fracture sets for which the fracture number is to be determined. By knowing the orientation of the fracture set and measuring the azimuth and plunge of the tape as it lies on the outcrop, one can calculate the angle between the line of the tape and the normal to the given fracture plane. This angle also can be measured from a stereographic net. The actual fracture number is:

$$F_n = \frac{F_t}{L} \cos \alpha \times 100 \qquad (9)$$

where F_t is the number of fractures in a given set which intersect the tape, L is the taped length, and α is the angle between the line of the tape and the normal to the fractures. The meaningful fracture number for a given lithology is found by averaging results from several similar stations. However, in analyzing the data one must make sure that there is no systematic change in, for example, structural position.

Before treating the bulk data, one should rotate the bedding strikes to a common strike line to determine the true nature of the fracture patterns. To make a proper interpretation, special attention should be given to stations where the entire pattern—two shear fractures and one extension fracture—occurs in a single outcrop, and these observations should strongly influence the interpretation of the bulk data. If data are gathered from many stations where the fractures all belong to patterns 1 and 2, the fractures will tend to group around the anticlinal dip and strike directions. Even though each grouping spreads ±30° from these directions, one may be tempted to interpret this as a single pattern of orthogonal fractures. This error can be avoided by referring to those stations where the total pattern is recorded in a single outcrop. If the beds at most of the stations dip more than about 5°, they should be rotated to a horizontal position before one makes the final interpretation. This is particularly true of fractures that strike at low angles to the bedding but dip at high angles to the bed.

Measurements from Aerial Photographs

The use of aerial photographs in fracture studies has advantages and disadvantages. The decision to use aerial photographs is, therefore, dictated by the final use of the data derived from them. One advantage is that strong fracture trends can be determined for large areas quickly and inexpensively. A disadvantage is that fractures are three-dimensional features, but any single surface trend is only a two dimensional representation and may represent several different fracture sets. Furthermore, no data concerning fracture morphology can be derived from air photos. If the desired end result involves only the surface trends of fractures—e.g., in secondary recovery programs—aerial photography can be a useful tool (Alpay, 1969). Aerial photographs should not be used without extensive ground control if the desired end result requires knowledge of: (1) fracture morphology, (2) small-scale fracture densities, (3) relative development of specific orientations of a conjugate pattern, or (4) determination of total fracture geometries.

Probably the most successful application of aerial photographs to fracture problems is a combined field and photo study. Initially, the major trends, changes in trend, and outcrop areas can be determined by measurements on the photomosaics. Specific locations for ground stations can be selected to sample properly the aerial control. Ground measurements can be made at these stations (as described in a preceding section), and any changes in attitude, relative development, morphology, *etc.*, can be integrated over the photographs from the surrounding ground control. This method not only saves time over studies that are totally ground-controlled, but it also makes fracture studies possible in areas with limited accessibility (because of topography or the attitude of landowners). However, even with this method caution must be exercised to avoid selection of anomalous ground stations which would have significant bearing in the analysis.

All complete surface studies should include examination of aerial photographs, because some fracture systems are too large to be observed on the ground. For example, the individual fractures of pattern 2 that make up the lineation in Figure 10 can be measured on the ground, but the fact that they form continuous zones can be seen only on the air photos.

SUMMARY AND CONCLUSIONS

A basis for the prediction of the relative development and orientations of fractures in an unknown province is provided by an understanding of the interactions between three primary factors:

1. The physical environment at the time of fracturing, *i.e.*, effective confining pressure, temperature, and strain rate. These three parameters strongly affect the behavior of rock material.
2. The magnitudes and orientations of the three principal stresses in the rock body at the time of fracture. The relative stress differences control the locations of the fractures, and the orientation of the stress field determines the potential fracture orientations.
3. The nature of the sedimentary layer, including the degree of induration at the time of fracturing and the relative thickness of the rock units. Thickness and lithology determine which rocks in a mixed sequence are more likely to be fractured.

Precise predictions of fracture spacing, areal extent, width, types (shear or extension), and exact locations within a rock body are not possible because the processes of fracturing and the compositions of the deformed body are so complex. It is possible to specify, at least approximately, the history of burial of a rock unit and something of its present structural geometry. From the former, the maximum temperature and overburden pressure affecting the rock can be determined. Qualitative predictions of the expected fracture development and orientation can be made from knowledge of the general geology, the known associations of fractures with structures, and the relations of fracturing to rock type, thickness, and structural position. Outcrop studies can be made initially and projected into the subsurface. As more seismic and well data become available, the prediction will become more quantitative. Though fracturing is a complicated process, laboratory and field studies provide as good a basis for estimating fracture development and trends as is available for many other geologic phenomena which are fearlessly predicted during the exploration of an area.

The recent literature reflects our growing ability to use fracture analysis as a productive geologic tool. The number of papers that pertain to laboratory techniques, field studies, and theory is increasing. The general relations of fracturing to lithology and geologic structure were summarized by Stearns (1967). Harris *et al.* (1960) described fracture development in the Big Horn basin, and this can be extended to any Rocky Mountain intermontane basin. In-

vestigations like those of Braunstein (1953), McCaleb and Willingham (1967), Murray (1968), Martin (1963), and Friedman (1969) illustrate how fracture problems can be treated in specific oil fields. McCaleb and Willingham demonstrated the importance of structural residual mapping and trend-surface analyses to increase production in fractured reservoirs. Murray applied quantitative methods to producing fields, and Friedman and Martin used subsurface data in a fracture analysis. The theoretical work of Hafner (1951) and Sanford (1959) showed the expected stress fields associated with certain types of faulting; from these the anticipated fracture patterns can be calculated. Other theoretical studies of folding—*i.e.*, Biot (1961) and Dieterich and Carter (1969)—consider not only how the stress trajectories change with time in the folding process, but also the delineation of the zones of highest strain where fracturing should be most intense. All these papers and many others have contributed to the knowledge of the three primary factors and to field application. There is now a firm basis from which to work, and fractured reservoirs should be found in the future—not by accident, but by intent.

REFERENCES

Adams, A. R., Ramey, H. J., Jr., and Burgess, R. J., 1968, Gas-well testing in a fractured carbonate reservoir: Jour. Petroleum Technology, v. 20, p. 1187–1194.
Alpay, A. O., 1969, Application of aerial photographic interpretation to the study of reservoir natural fracture systems: Soc. Petroleum Engineers, Ann. Fall Mtg., October, SPE No. 2567, 11 p.
American Geological Institute (J. V. Howell, chm.), 1957, Glossary of geology and related sciences: Natl. Research Council Pub. 501, 325 p.
Anderson, E. M., 1951, Dynamics of faulting and dyke formation, with applications to Britain, 2d ed.: Edinburgh, Oliver and Boyd, 206 p.
Anderson, T. O., and Stahl, E. J., 1967, A study of induced fracturing using an instrumental approach: Jour. Petroleum Technology, v. 19, p. 261–267.
Baker, W. J., 1955, Flow in fissured formations: 5th World Petroleum Cong. Proc., Sec. II/E, p. 379–393.
Barfield, E. C., Jordan, J. K., and Moore, W. D., 1959, An analysis of large scale flooding in the fractured Spraberry trend area reservoir: Jour. Petroleum Technology, v. 11, p. 15–19.
Biot, M. A., 1961, Theory of folding of stratified viscoelastic media and its implications in tectonics and orogenesis: Geol. Soc. America Bull., v. 72, p. 1595–1620.
Black, W. M., 1956, A review of drill-stem testing techniques and analysis: Jour. Petroleum Technology, v. 8, p. 21–30.
Blanchet, P. H., 1957, Development of fracture analysis as exploration method: Am. Assoc. Petroleum Geologists Bull., v. 41, p. 1748–1759.

Bloss, D. F., 1957, Anisotropy of fracture in quartz: Am. Jour. Sci., v. 225, p. 214–226.

Braunstein, Jules, 1953, Fracture-controlled production in Gilbertown field, Alabama: Am. Assoc. Petroleum Geologists Bull., v. 37, p. 245–249.

Briggs, R. O., 1964, Development of a downhole television camera: 5th Ann. SPWLA Logging Symposium, Midland, Texas, Houston, Texas, May; Soc. Prof. Well Log Analysts.

Cloos, Hans, 1936, Einführung in die Geologie: Berlin, Gebrüder Borntraeger, 503 p.

Coulomb, C. A. de, 1776, Sur une application des règles maximis et minimis à quelques problèmes de statique, relatif à l'architecture: Acad. Sci. Paris Mém. Math. Phys., v. 7, p. 343–382.

Daniel, E. J., 1954, Fractured reservoirs of Middle East: Am. Assoc. Petroleum Geologists Bull., v. 38, p. 774–815.

Dempsey, J. C., and Hickey, J. R., 1958, Use of a borehole camera for visual inspection of hydraulically-induced fractures: Producers Monthly, June, p. 18–21.

Dieterich, J. H., and Carter, N. L., 1969, Stress-history of folding: Am. Jour. Sci., v. 267, p. 129–154.

Doleschall, S., et al., 1967, The examination of the functional mechanism in the Nagylengyel-type fractured limestone reservoirs: Banyaszati Lapok, v. 100, p. 268–275.

Drummond, J. M., 1964, An appraisal of fracture porosity: Bull. Canadian Petroleum Geology, v. 12, p. 226–245.

Durham, C. A., Jr., 1962, Petroleum geology of Southeast Lincoln oil field, Kingfisher County, Oklahoma: Shale Shaker, v. 13, p. 2–10.

Dyes, A. B., and Johnston, O. C., 1953, Spraberry permeability from build-up curve analyses: AIME Trans., v. 198, p. 135–138.

Elkins, L. F., 1953, Reservoir performance and well spacing, Spraberry trend area field of West Texas: AIME Trans., v. 198, p. 177–196.

——— 1960, Determination of fracture orientation from pressure interference: AIME Trans., v. 219, p. 301–304.

Ells, G. D., 1962, Structures associated with the Albion-Scipio oil field trend: Michigan Geol. Survey Div., 86 p.

Emmett, W. R., Beaver, K. W., and McCaleb, J. A., 1969, Little Buffalo basin, Wyoming, Tensleep heterogeneity—its influence on infill drilling and secondary recovery: Soc. Petroleum Engineers Ann. Fall Mtg., October, SPE No. 2643, 8 p.

Fatt, I., 1953, The effect of overburden pressure on relative permeability: Jour. Petroleum Techology, v. 5, Tech. Note 194, October, p. 15–16.

——— 1958, Pore volume compressibilities of sandstone reservoir rock: Jour. Petroleum Technology, v. 10, Tech. Note 2004, Mar., p. 64–66.

——— and Davis, D. H., 1952, Reduction in permeability with overburden pressure: Jour. Petroleum Technology, v. 4, Tech. Note 147, Dec., p. 16.

Fraser, C. D., and Pettitt, B. E., 1962, Results of a field test to determine the type and orientation of a hydraulically induced formation fracture: Jour. Petroleum Technology, v. 14, p. 463–466.

Friedman, M., 1964, Petrofabric techniques for the determination of principal stress directions in rocks, p. 451–552 in State of stress in the earth's crust: New York, Elsevier, 732 p.

——— 1969, Structural analysis of fractures in cores from Saticoy field, Ventura County, California: Am. Assoc. Petroleum Geologists Bull., v. 53, p. 367–398.

Gardner, R. D., and Pincus, H. J., 1968, Fluorescent dye penetrants applied to rock fractures: Internat.

Jour. Rock Mechanics and Mining Sci., v. 5, p. 155–158.

Gray, K. E., 1962, How to plot pressure build-up curves: World Oil, v. 154, p. 82–91.

Griggs, D. T., and Handin, J. W., 1960, Observations of fracture and a hypothesis of earthquakes: Geol. Soc. America Mem. 79, p. 347–364.

Hafner, W., 1951, Stress distribution and faulting: Geol. Soc. America Bull., v. 62, p. 373–398.

Handin, J. W., 1966, Strength and ductility, Sec. 11 in Handbook of physical constants—revised edition: Geol. Soc. America Mem. 97, p. 223–289.

——— and Hager, R. V., 1957, Experimental deformation of sedimentary rocks under confining pressure—tests at room temperature on dry samples: Am. Assoc. Petroleum Geologists Bull., v. 41, p. 1–50.

——— and ——— 1958, Experimental deformation of sedimentary rocks under confining pressure—tests at high temperature: Am. Assoc. Petroleum Geologists Bull., v. 42, p. 2892–2934.

——— et al., 1963, Experimental deformation of sedimentary rocks under confining pressure: Pore pressure tests: Am. Assoc. Petroleum Geologists Bull., v. 47, p. 717–755.

Harp, L. J., 1966, Do not overlook fractured zones: World Oil, v. 162, p. 119–123.

Harris, J. F., Taylor, G. L., and Walper, J. L., 1960, Relation of deformational fractures in sedimentary rocks to regional and local structure: Am. Assoc. Petroleum Geologists Bull., v. 44, p. 1853–1873.

Hassebroek, W. E., and Waters, A. B., 1964, Advancements through 15 years of fracturing: Jour. Petroleum Technology, v. 4, p. 760–764.

Heard, H. C., 1962, The effect of large changes in strain rate in the experimental deformation of rocks: PhD thesis, Univ. California, Los Angeles.

Hodgson, R. A., 1961, Regional study of jointing in Comb Ridge–Navajo Mountain area, Arizona and Utah: Am. Assoc. Petroleum Geologists Bull., v. 45, p. 1–38.

Horner, D. R., 1951, Pressure build-up in wells: 3d World Petroleum Cong. Proc., Sec. II.

Hubbert, M. K., 1951, Mechanical basis for certain familiar geologic structures: Geol. Soc. America Bull., v. 62, p. 355–372.

——— and Willis, D. G., 1955, Important fractured reservoirs in the United States: 4th World Petroleum Cong. Proc., Sec. I/A/1, p. 58–81.

Huitt, J. L., 1956, Fluid flow in simulated fractures: Am. Inst. Chem. Engineers Jour., v. 2, p. 259–264.

Hunter, C. D., and Young, D. M., 1953, Relationship of natural gas occurrence and production in eastern Kentucky (Big Sandy gas field) to joints and fractures in Devonian bituminous shale: Am. Assoc. Petroleum Geologists Bull., v. 37, p. 282–299.

Husky, W. I., and Crawford, P. B., 1967, Performance of petroleum reservoirs containing vertical fractures in the matrix: Soc. Petroleum Engineers Jour., v. 7, p. 221–228.

Jenson, O. F., Jr., and Ray, William, 1965, Photographic evaluation of water wells: Log Analyst, March, p. 15–26.

Kafka, F. T., and Kirkbride, R. K., 1960, The Ragusa oil field, in Excursion in Sicily, May 27–30, Rome Petroleum Explor. Soc. Libya: p. 61–85.

Lamb, H., 1932, Hydrodynamics, 6th ed.: New York, Dover Publications, p. 581–582.

Malenfer, J., and Tillous, A., 1963, Study of the Hassi Messaoud field stratigraphy, structural aspect, study of detail of the reservoir: Inst. Franç. Pétrole Rev., v. 18, p. 851–867.

Martin, G. H., 1963, Petrofabric studies may find fracture porosity reservoirs: World Oil, v. 156, p. 52–54.

Matthews, C. S., and Russell, D. G., 1967, Pressure buildup and flow tests in wells: Soc. Petroleum Engineers Monograph Ser., no. 1.

McCaleb, J. A., and Wayhan, D. A., 1969, Geologic reservoir analysis, Mississippian Madison Formation, Elk Basin field, Wyoming-Montana: Am. Assoc. Petroleum Geologists Bull., v. 53, p. 2094–2113.

—— and Willingham, R. W., 1967, Influence of geologic heterogeneities on secondary recovery from Permian Phosphoria reservoir, Cottonwood Creek field, Wyoming: Am. Assoc. Petroleum Geologists Bull., v. 51, p. 2122–2132.

McGuire, W. J., and Sikora, V. J., 1960, The effect of vertical fractures on well productivity: AIME Trans., v. 219, p. 401–403.

Melton, F. A., 1929, A reconnaissance of the joint systems in the Ouachita Mountains and Central Plains of Oklahoma: Jour. Geology, v. 37, p. 733–738.

Miller, C. C., Dyes, A. B., and Hutchinson, C. A., Jr., 1950, The estimation of permeability and reservoir pressure from bottom hole pressure build-up characteristics: AIME Trans., v. 189, p. 91–104.

Morris, R. L., Grine, D. R., and Arkfeld, T. E., 1964, Using compressional and shear acoustic amplitudes for the location of fractures: Jour. Petroleum Technology, v. 4, p. 623–632.

Mullins, J. E., 1966, New tool takes photos in oil and mud-filled wells: World Oil, v. 163, p. 91–94.

Murray, G. H., Jr., 1968, Quantitative fracture study—Sanish pool, McKenzie County, North Dakota: Am. Assoc. Petroleum Geologists Bull., v. 52, p. 57–65.

Muskat, M., 1937, The flow of homogeneous fluids through porous media: New York, McGraw-Hill Book Co.,

Nechai, A. M., and Mel'nikov, D. A., 1963, A study of reservoir properties of strata in Northeast Caucasus areas by geophysical means: Vses. Nauchno-Issled. Inst. Geofiz. Metodov Razved. Trudy, p. 44–54 (in Russian).

Obert, L., and Stephenson, D. E., 1965, Stress conditions under which core discing occurs: Soc. Mining Engineers Trans., v. 232, p. 227–235.

O'Brien, C. A. E., 1953, Discussion of fractured reservoir subjects: Am. Assoc. Petroleum Geologists Bull., v. 37, p. 325–326.

Odeh, A. S., 1964, Unsteady-state behavior of naturally fractured reservoirs: Soc. Petroleum Engineers of AIME Proc., 39th Ann. Mtg., Houston, SPE Preprint No. 966, 6 p.

Ogle, B. A., 1961, Prospecting for commercial fractured "shale" reservoirs, Rocky Mountains: Am. Assoc. Petroleum Geologists Bull., v. 45, p. 407.

Pampe, C. F., 1963, Fort Trinidad's 25 sq. mile area promises big reserves: Oil and Gas Jour., v. 16, no. 27, p. 162–166.

Park, W. H., 1961, North Tejon oil field: California Div. Oil and Gas, California Oil Fields—Summ. Operations, v. 47, p. 13–22.

Parsons, R. W., 1966, Permeability of idealized fractured rock: Soc. Petroleum Engineers Jour., v. 6, p. 126–136.

Pickett, G. R., and Reynolds, E. B., 1969, Evaluation of fractured reservoirs: Soc. Petroleum Engineers Jour., v. 9, March, p. 28–38.

Pirson, S. J., 1953, Performance of fractured oil reservoirs: Am. Assoc. Petroleum Geologists Bull., v. 37, p. 232–244.

—— 1963, Handbook of well log analysis: Englewood Cliffs, Prentice-Hall, Inc., p. 303–314.

—— 1967, How to map fracture development from well logs: World Oil, v. 164, p. 106, 108, 113–114.

—— J. P. Trunz, Jr., and Gomez N., P., 1967, Fracture intensity mapping from well logs and from structure maps: 8th Ann. SPWLA Logging Symposium Trans., Denver; Houston, Texas, Soc. Prof. Well Log Analysts, p. B1–B23.

Pohly, R. A., 1962, Gravity work may aid search for Trenton fracture zones: World Oil, v. 154, p. 85–88.

Price, N. J., 1959, Mechanics of jointing in rocks: Geol. Mag., v. 96, p. 149–167.

Regan, L. J., 1953, Fractured shale reservoirs of California: Am. Assoc. Petroleum Geologists Bull., v. 37, p. 201–216.

—— and Hughes, A. W., 1949, Fractured reservoirs of Santa Maria district, California: Am. Assoc. Petroleum Geologists Bull., v. 33, p. 32–51.

Sanford, A. R., 1959, Analytical and experimental study of simple geologic structures: Geol. Soc. America Bull., v. 70, p. 19–51.

Shearrow, G. G., 1968, The story of Ohio's southeastern sleeper: Oil and Gas Jour., v. 66, p. 210–212.

Sheldon, Pearl, 1912, Observations and experiments on joint planes: Jour. Geology, v. 20, p. 54–79.

Smekhov, E. M., 1963, Fractured oil and gas reservoir problem and present status of its study: 2d U.N. Develop. Petroleum Resources, Asia and Far East Symposium Proc., v. 1, p. 476–481.

Snow, D. T., 1968, Rock fracture spacing openings and porosities: Am. Soc. Civil Engineers Proc., Jour. Soil Mechanics and Found. Div., v. 94, no. 1, p. 73–91.

Stearns, D. W., 1964, Macrofracture patterns on Teton anticline, northwest Montana (abs.): Am. Geophys. Union Trans., v. 45, p. 107–108.

—— 1967, Certain aspects of fracture in naturally deformed rocks, p. 97–118 in R. E. Riecker, ed., NSF advanced science seminar in rock mechanics: Bedford, Massachusetts, Air Force Cambridge Research Lab. Spec. Rept.

—— 1969, Fracture as a mechanism of flow in naturally deformed layered rocks: Conference on Research in Tectonics Proc.; Canada Geol. Survey Paper 68–52, p. 79–96.

Walters, R. F., 1953, Oil production from fractured Pre-Cambrian basement rocks in central Kansas: Am. Assoc. Petroleum Geologists Bull., v. 37, p. 300–313.

Ward, L. O., 1965, Mississippian Osage, northwest Oklahoma platform: Am. Assoc. Petroleum Geologists Bull., v. 49, p. 1562–1563.

Warren, J. E., and Root, P. J., 1963, The behavior of naturally fractured reservoirs: Soc. Petroleum Engineers Jour., v. 3, p. 245–255.

Wilkinson, W. M., 1953, Fracturing in Spraberry reservoir, W. Texas: Am. Assoc. Petroleum Geologists Bull., v. 37, p. 250–265.

Zemanek, J., et al., 1969, The Borehole Televiewer—a new logging concept for fracture location and other types of borehole inspection: Jour. Petroleum Technology, v. 246, p. 762–774.

Mechanics of Hydraulic Fracturing[1]

M. KING HUBBERT[2] and DAVID G. WILLIS[3]

Washington, D.C., and Atherton, California

Abstract A theoretical examination of the fracturing of rocks by means of pressure applied in boreholes leads to the conclusion that, regardless of whether the fracturing fluid is of the penetrating or nonpenetrating type, the fractures produced should be approximately perpendicular to the axis of least stress. The general state of stress underground is that in which the three principal stresses are unequal. For tectonically relaxed areas characterized by normal faulting, the least stress should be horizontal; the fractures produced should be vertical, and the injection pressure should be less than that of the overburden. In areas of active tectonic compression, the least stress should be vertical and equal to the pressure of the overburden; the fractures should be horizontal, and injection pressures should be equal to, or greater than, the pressure of the overburden.

Horizontal fractures cannot be produced by hydraulic pressures less than the total pressure of the overburden.

These conclusions are compatible with field experience in fracturing and with the results of laboratory experimentation.

INTRODUCTION

The hydraulic-fracturing technique of well stimulation is one of the major developments in petroleum engineering of the last decade. The technique was introduced to the petroleum industry in a paper by J. B. Clark of the Stanolind Oil and Gas Company in 1949. Since then, its use has progressively expanded so that, by the end of 1955, more than 100,000 individual treatments had been performed.

The technique itself is mechanically related to three other phenomena concerning which an extensive literature had previously developed. They are: (1) pressure parting in water-injection wells in secondary-recovery operations, (2) lost circulation during drilling, and (3) the breakdown of formations during squeeze-cementing operations—all of which appear to involve the formation of open fractures by pressure applied in a well bore. The most popular interpretation of this mechanism has been that the pressure had parted the formation along a bedding plane and lifted the overburden, notwithstanding the fact that, in the great majority of cases where pressures were known, they were significantly less than those due to the total weight of the overburden as determined from its density.

Prior to 1948, this prevalent opinion had already been queried by Dickey and Andresen (1945) in a study of pressure parting and by Walker (1946, 1949), who, in studies of squeeze cementing, pointed out that the pressures required were generally less than those of the overburden, and inferred that the fractures should be vertical. J. B. Clark (1949), in his paper introducing hydraulic fracturing, and later Howard and Fast (1950) and Scott et al. (1953), postulated that the entire weight of the overburden need not be lifted in producing horizontal fractures, but that it is only necessary to lift an "effective overburden," requiring a correspondingly lower pressure. Hubbert (1953a) pointed out that the normal state of stress underground is one of unequal principal stresses and that, in tectonically relaxed areas characterized by normal faults, the least stress should be horizontal. Therefore, in most cases, fracturing should be possible with pressure less than that of the overburden; moreover, such fractures should be vertical. Harrison et al. (1954), also on the expectation that the least principal stress should be horizontal, argued strongly in favor of vertical fracturing.

Scott et al. (1953) observed that, by the use of penetrating fluids, hollow, cylindrical cores could be ruptured at less than half the pressures required if nonpenetrating fluids were used. They also observed that, with penetrating fluids, the fractures occurred parallel with the bedding, irrespective of the orientation of the bedding with respect to the axis of the core; whereas, with nonpenetrating fluids, the fractures tended to be parallel with the axis of the core. Reynolds et al. (1954) described additional experiments confirming the earlier work of Scott et al. (1953) and concluded that it should be possible to produce horizontal fractures with penetrating fluids and vertical fractures with nonpenetrating fluids.

[1] Manuscript received, July 4, 1972. Published, in revised form, with permission of Society of Petroleum Engineers of AIME; originally published in Transactions of Society of Petroleum Engineers of AIME, 1957, v. 210, p. 153–168.

[2] U.S. Geological Survey.

[3] Consultant.

239

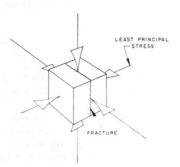

FIG. 1—Stress element and preferred plane of fracture.

During the last 2 years, the writers have been engaged in a critical reexamination of this problem, and, because the results obtained have sustained the conjecture offered earlier by Hubbert (1953), the principal content of this paper is an elaboration of that view.

STATE OF STRESS UNDERGROUND

The approach commonly made to the the problem of underground stresses is to assume that the stress field is hydrostatic or nearly hydrostatic, the three principal stresses being approximately equal to one another and to the pressure of the overburden. That this assumption cannot generally be true is apparent from the fact that, over long periods of geologic time, the earth has exhibited a high degree of mobility whereby the rocks have been repeatedly deformed to the limit of failure by faulting and folding. In order for such deformation to occur, substantial differences between the principal stresses are required.

The general stress condition underground is therefore one in which the three mutually perpendicular principal stresses are unequal. If fluid pressure were applied locally within rocks in this condition, and if the pressure were increased until rupture or parting of the rocks resulted, that plane along which fracture or parting would first be possible would be the one perpendicular to the least principal stress. It is here postulated that this plane is also the one along which parting is *most likely* to occur (Fig. 1).

In order, therefore, to have a mechanical basis for anticipating the fracture behavior of the rocks in various localities, it is necessary to know something concerning the stress states that can be expected. The best available evidence bearing upon these stress conditions is

the failure of the rocks themselves, either by faulting or by folding.

The manner in which the approximate state of stress accompanying various types of geologic deformation may be deduced was shown in a paper by Hubbert (1951); the rest of this section is a paraphrase of that paper.

Figures 2 and 3 show a box having a glass front and containing ordinary sand. In the middle, there is a partition which can be moved from left to right by turning a handscrew. The white lines are markers of powdered plaster of paris which have no mechanical significance. As the partition is moved to the right, a normal fault with a dip of about 60° develops in the left-hand compartment (Fig. 2). With further movement, a series of thrust faults with dips of about 30° develops in the right-hand compartment (Fig. 3).

The general nature of the stresses which accompany the failure of the sand can be seen in Figure 4. Adopting the usual convention of designating the greatest, intermediate, and least principal stresses by σ_1, σ_2, and σ_3, respectively (here taken as compressive), in the left-hand compartment σ_3 will be the horizontal stress, which is reduced as the partition is moved to

FIG. 2—Sand-box experiment showing normal fault.

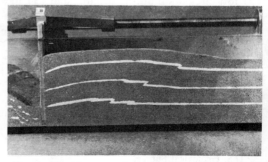

FIG. 3—Sand-box experiment showing thrust fault.

the right, and σ_1 will be the vertical stress, which is equal to the pressure of the overlying material. In the right-hand compartment, however, σ_1 will be horizontal, increasing as the partition is moved, and σ_3 will be vertical and equal to the pressure of the overlying material. A third type of failure, known as transcurrent faulting, is not demonstrated in the sandbox experiment. This type of failure occurs when the greatest and least principal stresses are both horizontal, and failure occurs by horizontal motion along a vertical plane. In all three kinds of faults, failure occurs at some critical relation between σ_1 and σ_3.

To determine this critical relation, it is first necessary to obtain an expression for the values of the normal stress σ and shear stress τ acting across a plane perpendicular to the $\sigma_1\sigma_3$-plane and making an arbitrary angle, α, with the direction of least principal stress, σ_3. As shown in Figure 5, these expressions may be obtained by balancing the equilibrium forces which act upon a small triangular prism of the sand. The resulting expressions for σ and τ are:

$$\sigma = \frac{\sigma_1 + \sigma_3}{2} + \frac{\sigma_1 + \sigma_3}{2}\cos 2\alpha, \text{ and}$$

$$\tau = \frac{\sigma_1 - \sigma_3}{2}\sin 2\alpha. \tag{1}$$

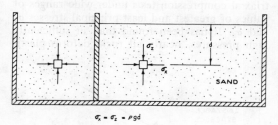

FIG. 4—Section showing approximate stress conditions in sand box.

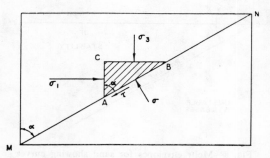

FIG. 5—Stresses σ and τ on plane of arbitrary angle α.

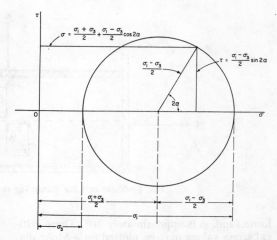

FIG. 6—Mohr diagram showing normal stress σ and shear stress τ on plane of orientation α in terms of σ_1, σ_3, and α.

A convenient method of graphically representing these expressions, known as the Mohr stress representation, consists of plotting values of normal and shear stress from equation 1 with respect to σ,τ coordinate axes for all possible values of the angle α, as shown in Figure 6. The locus of all σ,τ values is a circle; it can be seen that, as α approaches zero and the plane becomes normal to σ_1, the normal stress becomes equal to σ_1 and the shear stress disappears. However, as α approaches 90° and the plane becomes normal to the least principal stress, σ_3, the normal stress becomes equal to σ_3 and the shear stress again disappears. This figure completely describes all possible combinations of normal and shear stress acting on planes perpendicular to the plane containing σ_1 and σ_3.

It is next necessary to determine the combination of shear and normal stresses which will induce failure. This information can be obtained from a standard soil-mechanics test which is illustrated in Figure 7. A horizontally divided box is filled with sand, which is then placed under a vertical load. The shearing force which is necessary to displace the upper box is measured for various values of vertical stress. In this way, it is found that the shearing stress for failure is directly proportional to the normal stress, or that

$$\frac{\tau}{\sigma} = \tan\phi, \tag{2}$$

where ϕ is known as the angle of internal friction and is a characteristic of the material. For

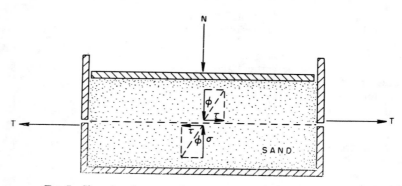

FIG. 7—Shear box for measuring ratio τ/σ at which slippage occurs.

loose sand, ϕ is approximately 30°. These critical stress values may be plotted on a Mohr diagram, as shown in Figure 8. The two diagonal lines comprise the Mohr envelopes of the material, and the area between them represents stable combinations of shear stress and normal stress, whereas the area exterior to the envelopes represents unstable conditions. Figure 8 thus indicates the stability region within which the permissible values of σ and τ are clearly defined. The stress circles may then be plotted in conjunction with the Mohr envelopes to determine the conditions of faulting. Such plots are illustrated in Figure 9 for both normal and thrust faulting. In both cases, one of the principal stresses will be equal to the overburden pressure, or σ_z. In the case of normal faulting, the horizontal principal stress is progressively reduced, thereby increasing the radius of the stress circle until it becomes tangent to the Mohr envelopes. At this point, unstable conditions of shear and normal stress are reached, and faulting occurs on a plane making an angle of 45° + $\phi/2$ with the least stress. For sand having an angle of internal friction of 30°, the normal fault would have a dip of 60°, which agrees with the previous experiments. For the case of thrust faulting, the least principal stress would be vertical and would remain equal to the overburden pressure, whereas the horizontal stress would increase progressively until unstable conditions occurred and faulting took place on a plane making an angle of 45° + $\phi/2$ with the least principal stress, or 45° − $\phi/2$ with the horizontal. For sand, this would be a dip of about 30°, which also agrees with the experiment.

It can be seen that, for sand having an angle of internal friction of 30°, failure will occur in either case when the greatest principal stress reaches a value which is about three times the least principal stress, and that the failure will occur along a plane making an angle of about 60° to the least principal stress. Also, for a fixed vertical stress σ_z, the horizontal stress may have any value between the extreme limits of one third and three times σ_z.

MOHR DIAGRAM FOR ROCKS

The foregoing theoretical analysis is directly applicable to solid rocks provided the Mohr envelopes have been experimentally determined. In order to make such determinations, it is necessary to subject rock specimens to a series of triaxial compression tests under wide ranges of values of greatest and least principal stresses σ_1 and σ_3. It has been found that, at sufficiently high pressures, nearly all rocks deform plasti-

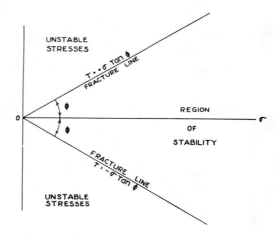

FIG. 8—Mohr envelopes for sand showing curves of values of σ and τ at which slippage occurs.

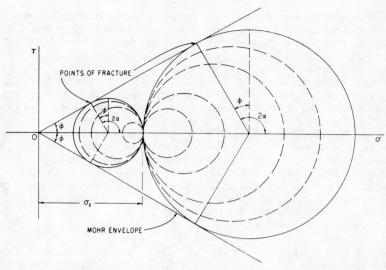

FIG. 9—Mohr diagram showing possible range of horizontal stress for a given vertical stress σ_z. Horizontal stress can have any value ranging from approximately one third vertical stress, corresponding to normal faulting, to approximately three times vertical stress, corresponding to reverse faulting.

cally and the Mohr envelopes become approximately parallel with the σ-axis. However, at lower pressures, most rocks fail by brittle fracture, and within this domain the envelopes are approximated by the equation

$$\tau = \pm (\tau_o + \sigma \tan \phi), \qquad (3)$$

where the angle of internal friction, ϕ, has values usually between 20° and 50°, and most commonly not far from 30°, and τ_0 is the shearing strength of the material for zero normal stress (Fig. 10).

Fortunately, equation 3 is applicable to most rocks within drillable depths. Exceptions would occur in the cases of the plastic behavior of rock salt and unconsolidated clays.

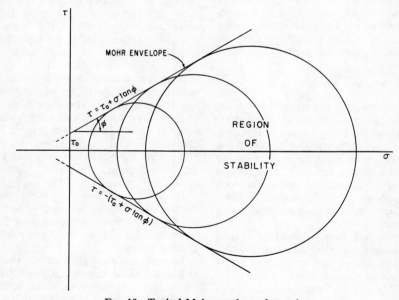

FIG. 10—Typical Mohr envelopes for rock.

One additional modification in the theoretical analysis is needed before it is directly applicable to geologic phenomena. Sedimentary rocks are both porous and permeable, and their pore spaces are almost invariably occupied by fluids, usually water, at some pressure, p. It is necessary to know the effect which is produced by the fluid pressure upon the mechanical properties of the rock.

This question was specifically investigated by McHenry (1948), who ran a large series of tests on duplicate specimens with and without enclosure by impermeable jackets, using nitrogen gas to produce the pressure (p). He found for the unjacketed specimens that, when the axial compressive stress, S, was corrected for the opposing fluid pressure, p, the value of the residual effective stress, $\sigma = S - p$, at which failure occurred was to a close approximation constant and independent of the pressure (p) of the permeating fluid.

This result is directly applicable to the behavior of rocks underground. Porous sedimentary rocks are normally saturated with fluid under pressure and constitute a mixed solid-fluid stress system. The stress field existing in this system may be divided into two partial stresses: (1) the hydrostatic pressure, p, which pervades both the fluid and solid constituents of the system, and (2) an additional stress in the solid constituent only. The total stress is the sum of these two.

If, across a plane of arbitrary orientation, S and T are the normal and tangential components, respectively, of the total stress, and σ and τ the corresponding components of the solid stress, then, by superposition,

$$S = \sigma + p, \text{ and}$$
$$T = \tau \qquad (4)$$

are the equations relating the stress fields.

The pressure (p) produces no shearing stress and hence has no tendency to cause deformation. Moreover, as demonstrated by the work of McHenry, it has no significant effect on the properties of the rock. Therefore, with respect to the stress of components σ and τ, the rock has the same properties underground as those exhibited in the triaxial testing machine using jacketed specimens.

This fact long has been recognized in soil mechanics, where the partial stress of components σ and τ is known as the effective stress and the pressure (p) as the neutral stress (Terzaghi, 1943). The effective stress defined in this manner is not to be confused with the postulated "effective overburden pressure" invoked by the Stanolind group and others to justify their assumption of horizontal fracturing in response to pressure less than that of the overburden.

Therefore, the entire Mohr stress analysis is directly applicable to porous rocks containing fluid under pressure, provided that only effective stresses are used.

It is interesting to consider the behavior of the effective vertical stress under various fluid-pressure conditions. Under any condition, the total vertical stress, S_z, is nearly equal to the weight of the overlying material per unit area. The effective vertical stress, σ_z, however, is given by

$$\sigma_z = S_z - p. \qquad (5)$$

Under normal hydrostatic conditions, p is somewhat less than half the total pressure of the overburden, and the effective vertical stress is therefore slightly more than half the overburden pressure. However, with abnormally high fluid pressures, such as occur in some parts of the Gulf Coast, the effective vertical stress is correspondingly reduced, and, in the extreme case of fluid pressure equal to the total overburden pressure, the effective vertical stress becomes zero.

In regard to the mechanical properties of rocks, for loosely consolidated sediments such as those of the Gulf Coast area, the limiting envelopes on the Mohr diagram will approximate those for loose sand shown in Figure 8. In older and stronger rocks, the Mohr envelopes are given approximately by equation 3. These envelopes are similar to those for loose sand except, as shown in Figure 10, they project to an intersection at some distance left of the origin, indicating that the rocks have some degree of tensile strength and also have a shear strength of finite magnitude τ_0 when the normal stress is equal to zero. The Mohr envelopes for tests on a sandstone and an anhydrite made by J. W. Handin of the Shell Development Exploration and Production Research Laboratory are shown in Figures 11 and 12.

In either case, however, it will be observed that, at other than shallow depths, the value of σ_3, the least stress, at the time of faulting will be of the order of one third the value of σ_1, the greatest stress.

Because these are the extreme states of stress at which failure occurs, it follows that the stress differences which prevail when actual faulting is not taking place are somewhat less than these limits. However, in most regions a certain type of deformation is usually repetitive over long geologic periods of time, indicating that the stresses of a given type are persistent and not far from the breaking point most of the time.

The orientation of the trajectories of the principal stresses in space is largely determined by the condition which they must satisfy at the surface of the earth. This is a surface along which no shear stresses can exist. For unequal stresses, the only planes on which the shear stresses are zero are those perpendicular to the principal stresses; therefore, one of the three trajectories of principal stress must terminate perpendicular to the surface of the ground, and the other two must be parallel with this surface. Thus, in regions of gentle topography and simple geologic structures, the principal stresses should be, respectively, nearly horizontal and vertical, with the vertical stress equal to the pressure of the overlying material.

Therefore, in geologic regions where normal faulting is taking place, the greatest stress, σ_1, should be approximately vertical and equal to the effective pressure of the overburden, whereas the least stress, σ_3, should be horizontal and most probably between one half and one third the effective pressure of the overburden.

However, in regions which are being shortened, either by folding or thrust faulting, the least stress should be vertical and equal to the effective pressure of the overburden, whereas the greatest stress should be horizontal and probably between two and three times the effective overburden pressure.

In regions of transcurrent faulting, both the greatest and least stresses should be horizontal, and the intermediate stress, σ_2, should equal the effective vertical stress.

As an example, the Tertiary sedimentary strata of the Texas and Louisiana Gulf Coast have undergone recurrent normal faulting throughout Tertiary time and to the present. Thus, a normal-fault stress system must have been continuously present, which intermittently reached the breaking points for the rocks, causing the stresses temporarily to relax and then gradually to build up again. Hence, during most of this time, including the present, a stress state must have existed in this region for which the least stress has been horizontal and probably between one half and one third of the effective pressure of the overburden. Because the faults in this area, except around salt domes, are mostly parallel with the strike of the rocks, the axis of least stress must be parallel with the dip.

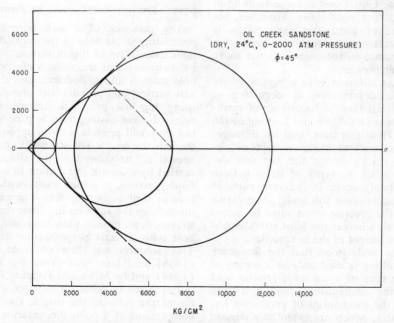

OIL CREEK SANDSTONE
(DRY, 24°C., 0–2000 ATM PRESSURE)
$\phi = 45°$

KG/CM^2

FIG. 11—Mohr envelopes for Oil Creek Sandstone (measurements by John Handin).

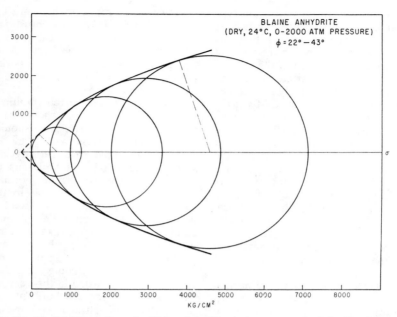

FIG. 12—Mohr envelopes for Blaine Anhydrite (measurements by John Handin).

A large part of the region of West Texas and the Mid-Continent is a region of tectonic relaxation characterized by older normal faults. The situation here is somewhat more ambiguous than that of the Gulf Coast because faulting in these regions is not now active. However, because evidence of horizontal compression is lacking, it is still reasonable to assume that a relaxed stress state in these areas is the more probable one at present.

In contrast, California is in a region where active tectonic deformation is occurring at present, as indicated by the recurrence of earthquakes, by extensive folding and faulting of the rocks during Holocene time, and by slippages along faults and measurable movements of elevation bench marks during the last few decades. All three of the types of stress pattern described probably occur in different parts of this region; but, in areas still undergoing active compression, the greatest stress must be essentially horizontal, whereas the least stress would be the effective weight of the overburden.

It should be understood that the foregoing analysis of faulting is used only as a means of estimating the state of stress underground, and that the shearing mechanism of faulting is quite distinct from the mechanism of producing hydraulic fractures, which are essentially tension phenomena. However, an understanding of the regional subsurface stresses makes it possible to analyze the stress conditions around the borehole and to determine the actual conditions under which hydraulic tension fractures will be formed.

STRESS DISTORTIONS CAUSED BY BOREHOLE

The presence of a well bore distorts the preexisting stress field in the rock. An approximate calculation of this distortion may be made by assuming that the rock is elastic, the borehole smooth and cylindrical, and the borehole axis vertical and parallel with one of the preexisting regional principal stresses. In general, none of these assumptions is precisely correct, but they will provide a close approximation to the actual stresses. The stresses to be calculated should all be viewed as the effective stresses carried by the rock in addition to a hydrostatic fluid pressure, p, which exists within the well bore as well as in the rock. The calculation is made from the solution in elastic theory for the stresses in an infinite plate containing a circular hole with its axis perpendicular to the plate. This solution was first obtained by Kirsch (1898) and later was given by Timoshenko (1934) and by Miles and Topping (1949).

Expressed in polar coordinates with the center of the hole as the origin, the plane-stress components at a point θ,r, exterior to the hole in a plate with an otherwise uniform uniaxial stress, σ_A, are given by

$$\sigma_r = \frac{\sigma_A}{2}\left[1 - \frac{a^2}{r^2}\right] + \frac{\sigma_A}{2}\left[1 + 3\frac{a^4}{r^4} - 4\frac{a^2}{r^2}\right]\cos 2\theta,$$

$$\sigma_\theta = \frac{\sigma_A}{2}\left[1 + \frac{a^2}{r^2}\right] - \frac{\sigma_A}{2}\left[1 + 3\frac{a^4}{r^4}\right]\cos 2\theta, \quad (6)$$

and

$$\tau_r{}^A = \frac{\sigma_A}{2}\left[1 - 3\frac{a^4}{I^4} + 2\frac{a^2}{r^2}\right]\sin 2\theta,$$

where a is the radius of the hole and the θ-axis is taken parallel with the axis of the compressive stress, σ_A. The same solution for a regional principal stress, σ_B, at right angles to σ_A, in which $\theta + 90°$ is used for the angular coordinate, may be superposed (added) onto equation 6 to give the complete horizontal components of stress in the vicinity of the borehole. The values of the horizontal stresses across the principal planes in the vicinity of the borehole have been calculated in this way for various relative values of the σ_B/σ_A ratio (Figs. 13, 14).

It can be seen that in every case the stress concentrations are local and that the stresses rapidly approach the undisturbed regional stresses within a few hole diameters. The principle of superposition of the two parts of the stress field is illustrated in Figure 13 for the case in which σ_B/σ_A is 1.0. For σ_A alone, the circumferential stress at the walls of the hole ranges from a minimum value of $-\sigma_A$ (tensile) across the plane parallel with the σ_A-axis to a maximum of $+3\sigma_A$ across the plane normal to the σ_A-axis. When the two stresses are superposed, the stress field has radial symmetry and the circumferential stress at the walls of the hole is $+2\sigma_A$. The resultant stress fields for other ratios of σ_B/σ_A are shown in Figure 14. In the extreme case, when $\sigma_B/\sigma_A = 3.0$, the circumferential stress at the walls of the hole ranges from a minimum of zero to a maximum of $+8\sigma_A$.

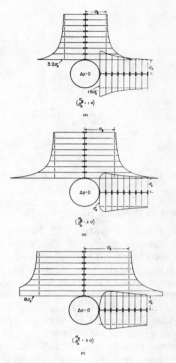

FIG 14—Stress states about a borehole for regional stress ratios σ_B/σ_A of 1.4, 2.0, 3.0

The vertical component of the stress is also distorted in the vicinity of the borehole. The initial vertical stress is equal to the effective pressure of the overburden. The distortion in the vertical stress is a function of the values of the regional horizontal stresses σ_A and σ_B. However, the magnitude of this distortion is small in comparison to the concentrations of the horizontal stresses, and it rapidly disappears with distance away from the well bore.

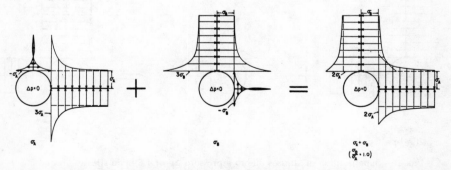

FIG. 13—Superposition of stress states about a well bore as a result of two horizontal principal stresses of equal magnitude.

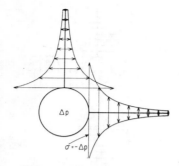

FIG. 15—Stresses caused by a pressure Δp within well bore.

Effect of Pressure Applied in Borehole

The application within the borehole of a fluid pressure in excess of the original fluid pressure produces additional stresses. For a nonpenetrating fluid, these stresses may be derived from the Lamé solution for the stresses in a thick-walled elastic cylinder, which was given by Timoshenko (1934). If the outer radius of the cylinder is allowed to become very large and the external pressure is set equal to zero, the solution becomes applicable to the well-bore problem, and the radial, circumferential, and vertical stresses become

$$\sigma_r = + \Delta p \frac{a^2}{r^2},$$

$$\sigma_\theta = - \Delta p \frac{a^2}{r^2}, \text{and} \qquad (7)$$

$$\sigma_z = 0,$$

in which Δp is the increase in fluid pressure in the well bore over the original pressure, a is the hole radius, and r is the distance from the center of the hole.

The circumferential stresses due to a pres-

sure Δp in the well bore are shown in Figure 15. The stresses given are those caused by Δp alone, and to obtain the complete stress field it is necessary to superpose these stresses upon those caused by the preexisting regional stresses which have been calculated. This method is illustrated in Figure 16, in which a pressure equal to 1.6 σ_A is applied to the well bore for the case in which $\sigma_B/\sigma_A = 1.4$ and is just sufficient to reduce the circumferential stress to zero across one vertical plane at the walls of the hole. In all cases, when the σ_B/σ_A ratio is greater than 1, the vertical plane across which σ_θ first becomes zero as the well-bore pressure is increased is that perpendicular to σ_A, the least horizontal stress.

Rupture pressures—In order to determine the rupture or breakdown pressures required to initiate fractures under various conditions, it is necessary to consider the properties of the rocks being fractured. The tensile strength of rock is a notoriously undependable quantity. For flawless specimens it ranges from zero for unconsolidated materials to several hundred pounds per square inch for the strongest rocks. However, as observation of any outcrop will demonstrate, flawless specimens of linear dimensions greater than a few feet rarely occur. In addition to the bedding laminations across which the tensile strength ordinarily is a minimum, the rocks usually are intersected by one or more systems of joints comprising partings with only slight normal displacements. Across these joint surfaces, the tensile strength is reduced essentially to zero.

In any section of a well bore a few tens of feet in length, it is probable that many such joints have been intersected. It appears likely, therefore, that the tensile strengths of most rocks that are to be subjected to hydraulic frac-

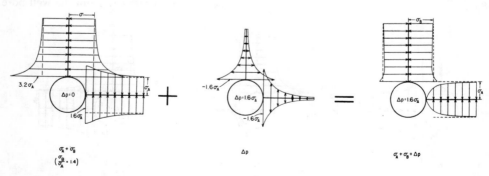

FIG. 16—Superposition of stresses due to a pressure Δp of 1.6 upon stresses around a well bore when σ_B/σ_A is 1.4.

turing by pressure applied in well bores are effectively zero, and that the pressure required to produce a parting in the rocks is only that required to reduce the compressive stresses across some plane in the walls of the hole to zero.

As the pressure is increased, the plane along which a fracture will commence will be that across which the compressive stress is first reduced to zero. In the case of a smooth cylindrical well bore, this plane must be vertical and perpendicular to the least principal regional stress. For the cases illustrated in Figure 14, the least compressive stress across a vertical plane at the walls of the hole ranges from twice σ_A to zero. Therefore, the down-the-hole pressure required to start a vertical fracture with a nonpenetrating fluid may vary from a value of twice the least horizontal regional stress to zero, depending upon the σ_B/σ_A ratio.

It can be seen from equation 7 that pressure inside a cylindrical hole in an infinite solid can produce no axial tension; thus, it is suggested that it is impossible to initiate horizontal fractures. However, under actual conditions in well bores, end effects should occur at well bottoms or in packed-off intervals in which axial forces equal to the pressure times the area of the cross section of the hole would be exerted upon the ends of the interval. Furthermore, irregularities exist in the walls of the borehole which should permit internal pressures to produce tension.

In particular, as has been suggested by Bugbee (1943), the initial fractures may be joints which have separated sufficiently to allow the entrance of the fluid, in which case it is only necessary to apply sufficient pressure to hold open and extend the fracture.

Injection pressures—Once a fracture has been started, the fluid penetrates the parting of the rocks and pressure is applied to the walls of the fracture, thereby reducing the stress concentration that previously existed in the vicinity of the well bore; the pressure, Δp, required to hold the fracture open in the case of a nonpenetrating fluid is then equal to the component of the undistorted stress field normal to the plane of the fracture. A pressure only slightly greater than this will extend the fracture indefinitely, provided it can be transmitted to the leading edge, as can be seen from an analysis of an ideally elastic solid (Fig. 17). The normal stresses across the plane of a fracture near its leading edge are shown for the case in which the applied pressure, Δp, is slightly greater than the original undistorted stress

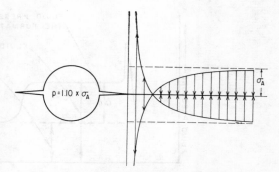

FIG. 17—Stresses in vicinity of a crack in a stressed elastic material when pressure acting on walls of crack is slightly greater than stress within the material.

field, σ_A. This solution is derived directly from the solution for the stresses in a semi-infinite solid produced by a distributed load, which was presented by Timoshenko (1934).

The tensile stress near the edge of the fracture approaches an infinite magnitude for a perfectly elastic material. For actual materials, this stress will still be so large that a pressure Δp only slightly greater than σ_A will extend the fracture indefinitely. The minimum down-the-hole injection pressure required to hold open and extend a fracture is therefore slightly in excess of the original undistorted regional stress normal to the plane of the fracture. The actual injection pressure will, in general, be higher than this minimum because of friction losses along the fracture.

Pressure behavior during treatment—A comparison of the breakdown and injection pressures required for nonpenetrating fluids and for various values of σ_B/σ_A shows that there are, in general, two types of possible down-the-hole pressure behavior during a fracturing treatment (Fig. 18). The pressures, Δp, are increases measured with respect to the original fluid pressure in the rocks. In one case the breakdown pressure might be substantially higher than the injection pressure, a situation which would probably correspond to a horizontal fracture from a relatively smooth well bore or to a vertical fracture under conditions in which the two horizontal principal stresses, σ_A and σ_B, were nearly equal. In the second case, there is no distinct pressure breakdown during the treatment, indicating that the pressure required to start the fracture is less than, or equal to, the injection pressure. Such a situation would correspond to a horizontal or vertical fracture starting from a preexisting open-

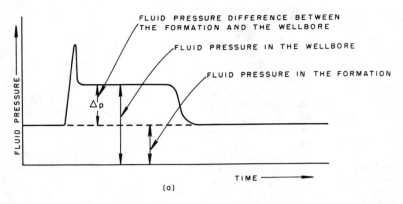

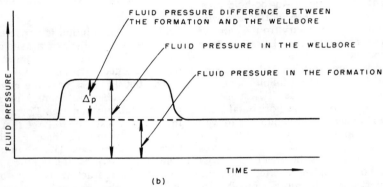

FIG. 18—Idealized diagram of two possible types of pressure behavior during fracture treatment depending upon various underground conditions.

ing or to a vertical fracture in a situation where the ratio σ_B/σ_A of the horizontal principal stresses was greater than 2.0.

Effect of Penetrating Fluids

When a penetrating fluid is used in a fracturing operation, a more complicated mechanical situation exists. As noted previously, the total normal stress S across any plane may be resolved into the sum of a residual solid stress σ and the fluid pressure p, or $S = \sigma + p$.

Furthermore, with a nonpenetrating fluid, an increase in pressure (Δp) equal to σ, or a total pressure $P = p + \Delta p$ equal to S, is required to hold open and extend a fracture along this plane.

For the case of a penetrating fluid, an increment of pressure in the fracture, which now may be designated by Δp_0, will produce an outward flow of fluid into the rock with a resulting variable increment Δp to the pressure within the formation. The gradient of this incremental pressure will exert an outward-directed force

upon both the rock and the contained fluid in each of the walls of the fracture. Let the normal component of this force acting upon the rock content of unit bulk volume be H_{n1} and that upon the fluid content, H_{n2}. Since the volume of the fluid per unit bulk volume is the porosity, f, and that of the rock $(1 - f)$, it follows that

$$H_{n1} = -(1 - f)\frac{\partial(\Delta p)}{\partial n}, \quad \text{and} \qquad (8)$$

$$H_{n2} = -f\frac{\partial(\Delta p)}{\partial n}. \qquad (9)$$

However, because of viscous coupling, the force H_{n2} acting upon the fluid is transmitted entirely to the rock, so that the total outward force exerted upon the rock per unit of bulk volume will be:

$$H_n = H_{n1} + H_{n2} = -\frac{\partial(\Delta p)}{\partial n}. \qquad (10)$$

Similarly, the total outward force per unit area of the fracture wall will be the integral of all the forces exerted upon the rock contained within a column of unit area of cross section normal to the fracture, or

$$\frac{F}{A} = -\int_0^\infty \frac{\partial(\Delta p)}{\partial n} dn = -\int_{\Delta p_0}^0 d(\Delta p) = \Delta p_0. \quad (11)$$

In order for the fracture to be held open and extended, this outward-directed force per unit area must be equal to σ. Therefore, for this case

$$P = p + \Delta p = p + \sigma = S, \quad (12)$$

which is exactly the same as the pressure required to hold the fracture open when a nonpenetrating fluid is used.

In the case of radial flow away from a well bore, the situation differs somewhat from that of flow away from a plane fracture. In the radial-flow case, a force acts outward whose magnitude per unit bulk volume is

$$\mathbf{H} = -\operatorname{grad}(\Delta p), \quad (13)$$

and the effect of this distributed field of force is to diminish the stress concentration at the face of the hole, in turn reducing the excess pressure that otherwise would be required to produce breakdown. Once the fracture is started, however, the flow field and the stress field become those associated with a plane fracture given in equation 12.

Therefore, the only effect of using a penetrating fluid is in the reduction of the breakdown pressure. The minimum injection pressure, for both penetrating and nonpenetrating fluids, must be greater than the preexisting normal stress across the plane of the fracture.

Orientation of Fractures Produced

Considering the earlier postulate that the fractures should occur along planes normal to the least principal stress, the minimum injection pressure should thus be equal to the least principal stress. Considering the injection pressures and fracture orientations for various tectonic conditions, it follows that, in regions characterized by active normal faulting, vertical fractures should be formed with injection pressures less than the overburden pressure; whereas, in regions characterized by active thrust faulting, horizontal fractures should be formed with injection pressures equal to, or greater than, the overburden pressure.

In the particular case of horizontal fracturing, the total normal stress across the plane of the fracture is equal to the pressure due to the total weight of the overburden; therefore, the minimum injection pressure, regardless of whether the fluid is penetrating or nonpenetrating, is also equal to the overburden pressure. It thus appears to be mechanically impossible for horizontal fractures to be produced with total fluid pressures less than the total overburden pressure.

Because the great majority of fracturing operations in the Gulf Coast, Mid-Continent, and West Texas–New Mexico regions require injection pressures less than the overburden pressure, it is difficult to escape the conclusion that most of these fractures are vertical. Furthermore, because the minimum pressures should be independent of the fluids used, there appears to be no valid basis for the claims that vertical versus horizontal fracturing can be controlled by variations in the penetrability of the fracturing fluids. In either case, it appears, the orientation of the fractures should be controlled by the preexisting stress field of the rocks into which the fluid is injected.

PREDICTED INJECTION PRESSURES

It is interesting to estimate the actual values of the minimum injection pressures under conditions of incipient normal faulting such as may exist in many parts of the Gulf Coast area.

As has been pointed out (equation 5), the undisturbed effective vertical stress σ_z is equal to the total pressure of the overburden S_z less the original fluid pressure p existing within the rocks prior to disturbances such as fluid withdrawals.

Under conditions of incipient normal faulting, the least principal stress, σ_A, will be horizontal and will have a value of approximately one third the effective overburden pressure, σ_z. Therefore,

$$\sigma_A \cong (S_z - p)/3. \quad (14)$$

Because the additional fluid pressure, Δp, required to hold open and extend a fracture should be equal to the least principal stress, then

$$\Delta p \cong (S_z - p)/3. \quad (15)$$

However, the total injection pressure, P, is given by

$$P = \Delta p + p. \quad (16)$$

Therefore,

$$P \cong (S_z + 2p)/3. \quad (17)$$

Dividing by depth z gives

$$P/z \cong (S_z/z + 2p/z)/3, \quad (18)$$

which is the approximate expression for the minimum injection pressure required per unit of depth in an area of incipient normal faulting.

The value of S_z/z is approximately equal to 1.00 psi per foot of depth for normal sedimentary rocks in most areas. Under normal hydrostatic fluid-pressure conditions, p/z is about 0.46 psi per foot of depth. Substituting these values into equation 18 gives

$$P/z \cong 0.64 \text{ psi/ft}$$

as the approximate minimum value that should be expected in the Gulf Coast.

Let us consider the values of P which would occur under conditions in which the original fluid pressure was other than hydrostatic. In those cases of an original fluid pressure less than hydrostatic, it can be seen from equation 17 that P would be correspondingly reduced. However, where abnormally high original fluid pressures prevail, P would become higher until, in the limit, when the original pressure p approaches the total overburden pressure S_z, P also approaches the total overburden pressure and fracturing will occur at pressures only slightly greater than the original fluid pressure.

Walker (1949) has described an interesting example of lost circulation which might be explained on the basis of the foregoing analysis. In a Gulf Coast well drilling below 10,000 ft (3,050 m), the specific weight of the drilling mud, which was a little over 18 lb/gal, had to be kept constant to within 0.3 lb/gal, or about 2 percent, to prevent either lost circulation when the density was too high or "kicking" by the formation fluids when the density was too low.

FIELD EVIDENCE

Present field data derived from experience with hydraulic fracturing, squeeze cementing, and lost circulation are fully consistent with the foregoing conclusions. In the Gulf Coast area, recent normal faulting indicates that vertical fractures should be formed with injection pressures less than the total overburden pressure. In the Mid-Continent and West Texas regions, old normal faulting, although representing more ambiguous evidence, also favors vertical fracturing.

Howard and Fast (1950) have summarized the pressure data from 161 squeeze-cementing and acidizing jobs performed in the Gulf Coast and West Texas–New Mexico areas. Also, published data by Harrison et al. (1954) and Scott et al. (1953) describe injection pressures for large samples from hydraulic-fracturing operations in the Gulf Coast, Mid-Continent, and West Texas regions. With but few exceptions, the injection pressures have been substantially less than the total overburden pressure, thus implying that vertical fractures are actually being formed.

In addition to the preceding data, the occurrence of lost circulation throughout the Gulf Coast area at pressure substantially less than that due to the weight of the overburden supports the conclusion that the least stress should be horizontal in this area.

In much of California, however, tectonic compression is taking place, and in these areas horizontal fractures should occur with injection pressures greater than the total overburden pressure. Although comparatively few fracturing operations have been performed in California, extremely high pressures are required with injection pressures commonly greater than the overburden pressure (W. E. Hassebroek, personal commun.).

A phenomenon very similar to artificial formation fracturing, but on a much larger scale, is that of dike emplacement. It has been pointed out by Anderson (1951) that igneous dikes should be injected along planes perpendicular to the axis of least principal stress. This situation is entirely analogous to that for artificial formation fracturing. A remarkable field example of the effect of a regional stress pattern upon the orientation of igneous dikes is the Spanish Peaks igneous complex in Colorado.

A map of this area is shown in Figure 19, and a photograph of West Spanish Peak from the northwest, showing dikes cutting flat-lying Eocene strata, is given in Figure 20. Odé (1957) has made a mathematical solution of the regional stress field which would most likely result from the presence of the structural features in the area. A comparison of the radial-dike system with the mathematical solution shows the dikes to be almost exactly perpendicular to the trajectories of the least principal stress.

EXPERIMENTAL FRACTURING DEMONSTRATION

In order to verify the inferences obtained theoretically, a series of simple laboratory experiments has been performed. The general procedure was to produce fractures on a small scale by injecting a "fracturing fluid" into a weak elastic solid which previously had been stressed. Ordinary gelatin (12-percent solution) was used for the solid, because it was sufficiently weak to fracture easily, was readily

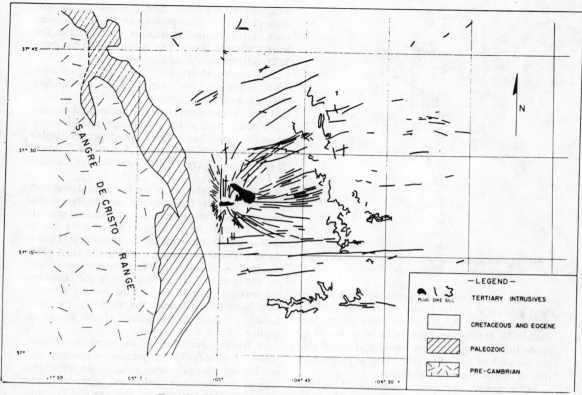

FIG. 19—Dike pattern of Spanish Peaks area, Colorado.

FIG. 20—Photograph of West Spanish Peak from northwest, showing dikes cutting flat-lying Eocene strata
(G. W. Stose, U.S. Geological Survey).

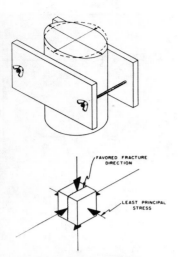

FIG. 21—Experimental arrangement for producing least stress in a horizontal direction.

molded with a simulated well bore, and was almost perfectly elastic under short-time application of stresses. A plaster-of-paris slurry was used as a fracturing fluid because it could be made thin enough to flow easily and could also be allowed to set, thus providing a permanent record of the fractures produced.

In a model experiment conducted in this way, the stress distributions are entirely independent of scale. Provided the material is elastic, similitude will exist no matter on what length scale the experiment is conducted.

The experimental arrangement consisted of a 2-gal polyethylene bottle with its top cut off, used as a container in which was placed a glass-tubing assembly consisting of an inner mold and concentric outer casings. The container was sufficiently flexible to transmit externally applied stresses to the gelatin. The procedure was to place the glass-tubing assembly in the liquid gelatin and, after solidification, to withdraw the inner mold leaving a "well bore" cased above and below an open-hole section. Stresses were then applied to the gelatin in two ways. The first method (Fig. 21) was to squeeze the polyethylene container laterally, thereby forcing it into an elliptical cross section and producing a compression in one horizontal direction and an extension at right angles in the other. The least principal stress was therefore horizontal, and vertical fractures would be expected in a vertical plane, as shown in Figure 21.

In other experiments the container was wrapped with rubber tubing stretched in tension (Fig. 22), thus producing radial compression and a vertical extension. In this case, the least principal stress was vertical, and horizontal fractures would be expected.

The plaster slurry was injected from an aspirator bottle to which air pressure was applied by means of a squeeze bulb.

Four experiments were performed under each of the two stress conditions, and in every case the fractures were formed perpendicular to the least principal stress. A vertical fracture is shown in Figure 23 and a horizontal fracture in Figure 24.

The saucer shape of the horizontal fracture is a result of the method of applying the stresses. As the gelatin is compressed on all sides, it tends to be displaced vertically but is restrained by the walls of the container. Thus a shear stress is produced, causing the least principal stress to intersect the container at an angle from above. Therefore, when the fractures are formed normal to the least principal stress, they turn upward near the walls of the container, producing the saucer shape shown in Figure 24.

A further variation in the experiment consisted of stratifying the gelatin by pouring and solidifying alternate strong and weak solutions. One experiment was performed in this way under each stress condition. The vertical fracture is illustrated in Figure 25, in which the weak gelatin appeared to fracture slightly more read-

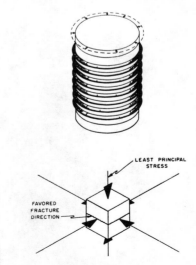

FIG. 22—Experimental arrangement for producing least stress in a vertical direction.

ily than the strong gelatin. Figure 26 shows a horizontal fracture in stratified gelatin. In this case, the fracture is not saucer shaped but appears to have followed a plane of weakness created by bubbles between two gelatin layers.

SIGNIFICANCE OF VERTICAL FRACTURING IN RESERVOIR ENGINEERING

In view of the foregoing evidence, it now appears fairly definite that most of the fracturing produced hydraulically is vertical rather than horizontal, so the significance of this fact in reservoir engineering should be mentioned. In geologically simple and tectonically relaxed areas, the regional stresses should be fairly uniform over extensive areas so that the horizontal stress trajectories in local areas should be nearly rectilinear. Consequently, when numerous wells in a single oil field are fractured, the fractures should be collimated by the stress field to almost the same strike.

There are serious implications, as Crawford and Collins (1954) have pointed out, with respect to the direction of drive and the sweep efficiency in secondary-recovery operations. If the direction of drive should be parallel with the strike of the fractures, then the flow would

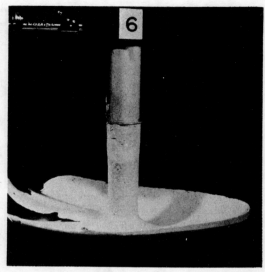

FIG. 24—Horizontal fracture produced under stress conditions illustrated in Figure 22.

be effectively short-circuited and the sweep efficiency would be very low. However, if the drive were normal to the strike of the fractures, the flow pattern would approximate that between parallel line sources and sinks and the sweep efficiency would approach unity.

This circumstance emphasizes the need, which is becoming increasingly more urgent, for the development of reliable downhole instruments by means of which not only the ver-

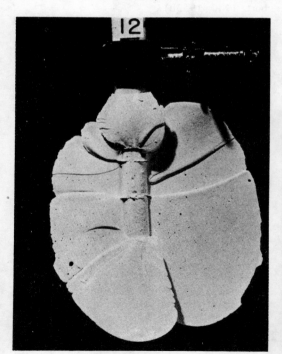

FIG. 23—Vertical fracture produced under stress conditions illustrated in Figure 21.

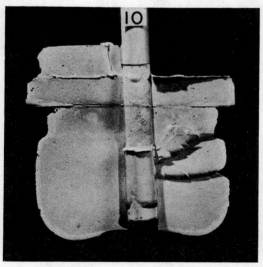

FIG. 25—Vertical fracture in stratified gelatin.

tical extent, but also the azimuth of the fractures, can be determined.

Since the foregoing paragraphs were written, these theoretical inferences have been strikingly confirmed by the fracturing experience during waterflood operations of the North Burbank field, Oklahoma. According to Z. Z. Hunter (1956), the initial pilot flood was based on the conventional five-spot pattern of injection and producing wells, but the results were anomalous. The injection wells were broken down at very low pressures (as low as one fourth of the overburden pressure), and producing wells east and west of injection wells were commonly bypassed by the flood. Finally, a sudden influx of water occurred in the isolated "Stanley Stringer" sandstone (in the "Burbank" sandstone) 1 mi (1.6 km) east of the flood area.

Cumulative experiences of this kind, supplemented by fracture observations in oriented cores, led to the conclusion that the fractures were essentially vertical and oriented east and west. This realization led to a change of procedure wherein line drives were instituted from east-west rows of fractured injection wells to alternate rows of fractured producing wells. Greatly increased oil production without a corresponding increase in the water-oil ratio resulted.

Another question to be considered concerns the vertical migration of fluids. It is obvious that vertical fractures will facilitate the vertical migration of fluids where the fractures intersect permeability barriers. They may in this manner interconnect several separate reservoirs in lenticular sandstones imbedded in shales, and may in fact tap some such reservoirs not otherwise in communication with the fractured well. There is a danger, however, where a reservoir is overlain by a thin permeability barrier and a water-bearing sand (or sandstone), that a vertical fracture may also permit the escape of the oil and gas into the barren sands (or sandstones) above.

A related question is that of the effect on water production of a vertical fracture which extends across the oil-water interface. In order to obtain an approximate idea of what this effect may be, consider a reservoir composed of a thick sand which is homogeneous and isotropic with respect to permeability. If production prior to fracturing is from an interval well above the water table, the water will form a radially symmetrical cone, with a slope whose sine at any point is given by

$$\sin \theta = - \frac{\rho_0}{\rho_w - \rho_0} \cdot \frac{1}{g} \cdot \left| \text{grad } \Phi_0 \right|, \qquad (19)$$

where ρ_0 and ρ_w are the densities of oil and water, respectively, g is the acceleration of gravity, and ϕ_0 is the potential of oil (Hubbert, 1940, 1953b).

The oil potential ϕ_0 at a given point is defined by

$$\Phi_0 = gz + \frac{p}{\rho_0}, \qquad (20)$$

where z is the elevation of the point with respect to sea level and p is the gauge pressure. Then, by Darcy's law, the volume of fluid crossing a unit area in unit time will be

$$q = - \frac{k\rho_0}{\mu} \text{ grad } \Phi_0, \qquad (21)$$

where k is the permeability of the sand and μ is the fluid viscosity. Substitution into equation 19 gives, for the tilt of the oil-water interface,

$$\sin \theta = \frac{\rho_0}{\rho_w - \rho_0} \cdot \frac{\mu}{gk\rho_0} \cdot q. \qquad (22)$$

Hence, the sine of the angle of tilt is proportional to the rate of flow, q, of the oil along the interface.

We have now only to consider the flow patterns about the well without and with vertical fracturing. Without fracturing the flow converges radially toward the well with a rapidly increasing flow rate and a corresponding steepening of the cone. With fracturing, for the same rate of oil production from the well, the flow pattern approximates that of linear flow toward a vertical-plane sink. The maximum values of the flow velocity, q, for this case will be very much less than for the radial-flow case. Hence, for a given rate of oil production, a vertical fracture across the oil-water interface in a uniform sandstone, instead of causing an increase of water production, actually should serve to reduce markedly the water coning and, consequently, to decrease the production of water—a result in accord with reports of field experience wherein fracturing near the water table has not resulted in increased water production.

CONCLUSIONS

From the foregoing analysis of the problem of hydraulic fracturing of wells, the following general conclusions appear to be warranted.

1. The state of stress underground is not, in general, hydrostatic, but depends upon tectonic conditions. In tectonically relaxed areas characterized by normal faulting, the least stress will be approximately horizontal; whereas, in areas of tectonic compression characterized by folding and thrust faulting, the least stress will be approximately vertical and, provided the deformation is not too great, approximately equal to the overburden pressure.

2. Hydraulically induced fractures should be formed approximately perpendicular to the least principal stress. Therefore, in tectonically relaxed areas, they should be vertical, whereas, in tectonically compressed areas, they should be horizontal.

3. Rupture or breakdown pressures are affected by the values of the preexisting regional stresses, by the hole geometry including any preexisting fissures, and by the penetrating quality of the fluid.

4. Minimum injection pressures depend solely upon the magnitude of the least principal regional stress and are not affected by the hole geometry or the penetrating quality of the fluid. In tectonically relaxed areas, the fractures should be vertical and should be formed with injection pressures less than the total overburden pressure. In tectonically compressed areas, provided the deformation is not too great, the fractures should be horizontal and should require injection pressures equal to, or greater than, the total overburden pressures.

5. It does not appear to be mechanically possible for horizontal fractures to be produced in relatively undeformed rocks by means of total injection pressures which are less than the total pressure of the overburden.

6. In geologically simple and tectonically relaxed areas, not only should the fractures in a single field be vertical, but they also should have roughly the same direction of strike.

7. Vertical fractures intersecting horizontal permeability barriers will facilitate the vertical flow of fluids. However, in the absence of such barriers, vertical fractures across the oil-water or gas-oil interface will tend to reduce the coning of water or gas into the oil section for a given rate of oil production.

REFERENCES

Anderson, E. M., 1951, The dynamics of faulting and dyke formation with applications to Britain: Edinburgh and London, Oliver and Boyd, 2d ed.

Bugbee, J. M., 1943, Reservoir analysis and geologic structure: AIME Trans., v. 151, p. 99.

Clark, J. B., 1949, A hydraulic process for increasing the productivity of wells: AIME Trans., v. 186, p. 1–8.

Crawford, P. B., and R. E. Collins, 1954, Estimated effect of vertical fractures on secondary recovery: AIME Trans., v. 201, p. 192.

Dickey, P. A., and K. H. Andresen, 1945, The behavior of water-input wells: Drilling and Prod. Practices, v. 34.

Harrison, Eugene, W. F. Kieschnick, Jr., and W. J. McGuire, 1954, The mechanics of fracture induction and extension: AIME Trans., v. 201, p. 252.

Howard, G. C., and C. R. Fast, 1950, Squeeze cementing operations: AIME Trans., v. 189, p. 53.

Hubbert, M. K., 1940, The theory of ground-water motion: Jour. Geology, v. 48, p. 785–944.

——— 1951, Mechanical basis for certain familiar geologic structures: Geol. Soc. America Bull., v. 62, no. 4, p. 355–372.

——— 1953a, Discussion of paper by Scott, Bearden, and Howard, "Rock rupture as affected by fluid properties": AIME Trans., v. 198, p. 122.

——— 1953b, Entrapment of petroleum under hydrodynamic conditions: Am. Assoc. Petroleum Geologists Bull., v. 37, no. 8, p. 1954–2026.

Hunter, Z. Z., 1956, 8½ Million extra barrels in 6 years: Oil and Gas Jour., August 27, p. 86.

Kirsch, G., 1898, Die Theorie der Elastizität und die Bedürfnisse der Festigkeitslehre: Zeitschr. Ver. Deutsch. Ingenieure, v. 42, p. 797.

McHenry, Douglas, 1948, The effect of uplift pressure on the shearing strength of concrete: Troisieme Congres des Grands Barrages, Stockholm, Sweden.

Miles, A. J., and A. D. Topping, 1949, Stresses around a deep well: AIME Trans., v. 179, p. 186

Odé, H., 1957, Mechanical analysis of the dike pattern of the Spanish Peaks area, Colorado: Geol. Soc. America Bull., v. 68, no. 5, p. 567–575.

Reynolds, J. J., P. E. Bocquet, and R. C. Clark, Jr., 1954, A method of creating vertical hydraulic fractures: Drilling and Prod. Practice, p. 206.

Scott, P. P., Jr., W. G. Bearden, and G. C. Howard, 1953, Rock rupture as affected by fluid properties: AIME Trans., v. 198, p. 111.

Terzaghi, Karl, 1943, Theoretical soil mechanics: New York, John Wiley and Sons.

Timoshenko, S., 1934, Theory of elasticity: New York and London, McGraw-Hill.

Walker, A. W., 1946, Discussion of paper by A. J. Teplitz and W. E. Hassebroek, "An investigation of oil-well cementing": Drilling and Prod. Practice, p. 102.

——— 1949, Squeeze cementing: World Oil, September, p. 87.

THE AMERICAN ASSOCIATION OF PETROLEUM GEOLOGISTS BULLETIN
Vol. 57, No. 9, September, 1973

Subsurface Petrographic Study of Joints in Variegated Siltstone-Sandstone and Khairabad Limestone, Pakistan[1]

S. M. SHUAIB[2]

North Nazimabad, Karachi-33, Pakistan

Abstract Petrographic methods for the determination of joint parameters such as volume density of joints, and joint porosity and permeability have been applied to selected core samples from the Nuryal 1 well drilled in the Potwar region of West Pakistan.

Joint porosity and permeability data from Jurassic variegated siltstone-sandstone and Paleocene Khairabad Limestone from selected core sections of the Nuryal 1 well at depths between 4,662 and 4,751 m indicate the presence of fairly effective reservoir potential in Jurassic variegated siltstone-sandstone and effective reservoir beds in the Paleocene Khairabad Limestone.

INTRODUCTION

Study of rock joints in the petrographic section of the geologic and analytical laboratories of the Oil and Gas Development Corporation, Pakistan, was initiated in 1971 by consultant chief geologist L. K. Teplov and consultant on reservoir properties of rocks, Yu. A. Bourlakov. This paper is based on the method and procedure used by the Russian geologists since 1955 to determine joint parameters such as volume density of joints, joint porosity, joint permeability, *etc.*, on fissured reservoirs.

Oil and gas reservoirs are divided into two main types—granular and joint. Granular reservoirs generally are silty-sandy rocks characterized by intergranular porosity. Some limestone and dolomite with oolitic, intercrystalline, and intergranular porosity may be classified as reservoirs of granular type. The joint type of reservoirs can be almost any rigid rock type

[1] Manuscript received, November 20, 1972; accepted, January 6, 1973.

[2] Oil and Gas Development Corporation.

The writer is indebted to Dr. M. H. Khan, Chief of Laboratories, for providing this interesting problem and necessary facilities. Thanks are due also to ex-consultant chief geologist L. K. Teplov, ex-consultant on reservoir properties of rocks Yu. A. Bourlakov, and senior petrologist Dr. I. R. Beg for helpful discussion and cooperation.

which has undergone the effects of tectonic stresses, weathering, leaching, crystallization, *etc.*, and which is broken by joints.

The object of the comprehensive study of fissured reservoirs was to determine the nature of joints, direction and magnitude of tectonic forces, effect on the porosity and permeability, and rock correlations. Oil and gas fields are related to fissured reservoirs. Despite the fact that the reservoir rocks have low intergranular permeability, they can yield high-flow rates of oil and gas because of the presence of joints. Methods for the determination of joints are geologic, geophysical, and hydrodynamic. This paper deals only with the petrographic study of joints in large thin sections from selected core samples of Nuryal 1 well, drilled near the village of Dhok Nuryal in the Potwar region of West Pakistan. Selected Jurassic variegated siltstone-sandstone and Paleocene Khairabad Limestone from cores 23 to 28 at depths between 4,662 and 4,751 m were analyzed petrographically and the results are shown in Table 1 and Figure 1.

METHOD

Slices were cut perpendicular to the bedding planes in selected core samples at different intervals according to the changes in lithology and the number of joints. Large thin sections were made from these slices for the following petrographic study of joints.

Nature of joints—Joints may be horizontal, vertical, chaotic, *etc*. They may be filled partly or completely by mineral constituents, bitumens, or both. Some joints may not be filled by any constituent and will remain open. Joints filled by mineral constituents are not considered here; only open and bitumen-filled joints have been measured for the calculation of effective porosity and permeability. The average width of microjoints was calculated by tak-

Table 1. Jointing Parameters of Rocks from Nuryal Well 1

Core No.	Depth (M)	Formation and Age	System of Jointing	Area of Slide in mm^2 (S)	Length of Joints in mm (l)	Aver. Width (mm) (b)	Volume Density per Meter (T)	Jointing Porosity (%) (m$_T$)	Jointing Permeability (md)	Brief Petrography
23 (top) to (bottom)	4,662 4,677	Khairabad Limestone (Paleocene)	Horizontal Chaotic	35X15 35X20	45 45	0.04 0.01	135 101	0.345 0.064	187.6 1.1	Limestone, dark-gray, fossiliferous Limestone, gray, fossiliferous
24 (top) (top) bottom) bottom)	4,690 4,703	Khairabad Limestone (Paleocene)	Chaotic Chaotic Chaotic Chaotic	50X25 50X35 40X40 45X35	140 135 300 200	0.04 0.04 0.04 0.04	176 121 295 200	0.448 0.309 0.750 0.508	121.8 84.4 205.2 139.0	Limestone, dark-gray to gray, fossil. Limestone, dark-gray to gray, fossil. Limestone, dark-gray, fossiliferous Limestone, dark-gray, fossiliferous
25 (top) (top) (middle) (middle) (bottom) (bottom)	4,703 4,720	Khairabad Limestone (Paleocene)	Chaotic Chaotic Chaotic Chaotic Chaotic Chaotic	50X30 45X32 20X20 20X25 40X25 40X25	160 155 10 25 25 25	0.04 0.04 0.02 0.01 0.04 0.04	168 169 39 79 39 39	0.427 0.431 0.050 0.050 0.100 0.100	116.7 117.8 3.4 0.9 27.4 27.4	Limestone, dark-gray to gray, fossil. Limestone, dark-gray to gray, fossil. Limestone, gray, fossiliferous Limestone, gray, fossiliferous Limestone, gray, fossiliferous Limestone, gray, fossiliferous
26 (top) to	4,731 4,734	Variegated series (Jurassic)	Horizontal	45X40	110	0.02	96	0.122	16.7	Siltstone, gray, noncalcareous, carbonaceous, argillaceous
27 (top) to	4,740 4,742	Variegated series (Jurassic)	Horizontal	50X35	100	0.02	90	0.114	15.6	Sandstone, light- to dark-gray, non-calcareous, argillaceous, silty, very fine to fine-grained
28 (top) to	4,742 4,751	Variegated series (Jurassic)	Chaotic	50X35	120	0.03	108	0.206	31.7	Sandstone, whitish-gray, noncalcareous, carbonaceous, fine-grained, quartzose

ing the mean value from several measurements of the thickness of the open or bitumen-filled joints at approximate intervals. Only a few of these exceed a width of 0.1 mm. Joints with a width of 0.001 to 0.01 mm are considered narrow; 0.01 to 0.05 mm, medium; and 0.05 to 0.1 mm, wide.

Volume density of joints—Volumetric density of joints is a criterion of rock fracturing, which is calculated from the formula:

$$T = \frac{\pi \times l}{2S},$$

where T is volume density of joints, $\pi = 3.14$, l is the total length of joints present, and S is the area of rock section. T preferably is calculated per meter.

A volumetric density of joints less than 50 per meter is considered as low, 50–100 as moderate, and greater than 100 as high.

Joint porosity—The total porosity of jointed rocks consists of intergranular porosity, porosity caused by the presence of open or bitumen-filled joints, and the volume of voids present. Fractured limestone commonly has voids of different dimensions. Voids are considered as very fine if the diameter is between 0.01–0.1 mm; fine if 0.1–0.25 mm; medium if 0.25–0.50 mm; large if 0.5–1.0 mm; and coarse if 1.0–

2.0 mm. Voids having diameters greater than 2.0 mm are considered caverns.

The figure given in this paper is joint porosity, which in actual fact is a very small part of the total porosity of jointed rocks and rarely exceeds 1 percent. The formula for the calculation of joint porosity is:

$$m_T = \frac{b \times l}{S} \times 100,$$

where m_T represents joint porosity in percent, b is average width of open and bitumen-filled joints in millimeters, l is the total length of joints in millimeters, and S is the area of rock section in square millimeters.

Joint porosity less than 0.1 percent is considered as low; 0.1–0.5 percent, medium; and greater than 0.5 percent, high.

Joint permeability—The total permeability of jointed rocks is determined by the permeability of intergranular space and permeability of joints intersecting within rocks. Fissured reservoirs invariably are almost connected, even in very compact and brittle rocks where most intergranular permeability does not exceed 0.1 md. In such rocks, the main permeability is caused by the presence of joints. Thus the determination of joint permeability is important and is calculated from the formula:

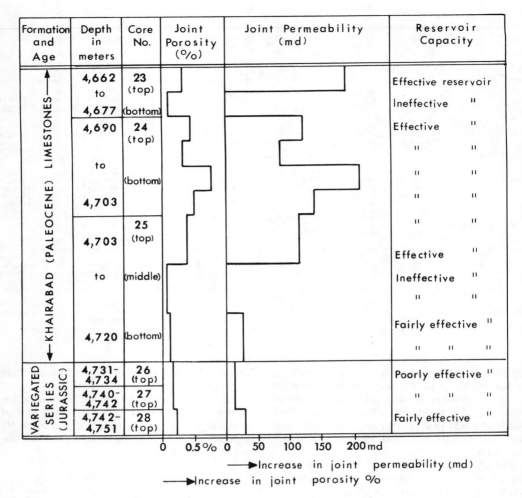

Formation and Age	Depth in meters	Core No.	Joint Porosity (%)	Joint Permeability (md)	Reservoir Capacity
KHAIRABAD (PALEOCENE) LIMESTONES	4,662 to 4,677	23 (top) (bottom)			Effective reservoir / Ineffective "
	4,690 to 4,703	24 (top) (bottom)			Effective " / " " / " " / " "
	4,703 to 4,720	25 (top) (middle) (bottom)			" " / Effective " / Ineffective " / " " / Fairly effective " / " " "
VARIEGATED SERIES (JURASSIC)	4,731–4,734	26 (top)			Poorly effective "
	4,740–4,742	27 (top)			" " "
	4,742–4,751	28 (top)			Fairly effective "

0   0.5 % 0   50   100   150   200 md

→ Increase in joint permeability (md)

→ Increase in joint porosity %

Horizontal scale for joint porosity: 0–100md=1 %

No vertical scale

FIG. 1—Representation of joint porosity, permeability, and reservoir capacity of Khairabad Limestone and Jurassic. Variegated series from Nuryal well 1, Potwar, West Pakistan.

$$K_T = \frac{A \times B \times b^3 \times l}{S},$$

where K_T is joint permeability in millidarcys; A is a constant for the same system of joints (its value is equal to 3.4×10^6 for one system of horizontal bedding joints, 2.28×10^6 for three reciprocally perpendicular systems of joints, 1.71×10^6 for two reciprocally perpendicular systems of vertical joints and also for chaotically arranged joints); $B = 10$ if permeability is calculated in millidarcys; b is the average width of open and bitumen-filled joints in millimeters;

l is the total length of joints in millimeters; and S is the area of rock section in square millimeters.

Rocks having joint permeability less than 5 md are considered to be ineffective reservoirs; 5–25 md as poorly effective reservoirs; 25–50 md as fairly effective reservoirs, and greater than 50 md as effective reservoirs.

DISCUSSION

The Jurassic variegated siltstone-sandstone of the Potwar region is light- to dark-gray and whitish-gray, medium hard to hard, moderately compact to compact, noncalcareous, carbonaceous, somewhat fractured, argillaceous siltstone to silty fine-grained sandstone with detrital sandy grains (mainly quartz). Microjoints in thin sections show both horizontal and chaotic arrangements of joints, which are filled by argillaceous or carbonaceous materials or both. The average width of open or carbonaceous-filled joints ranges from 0.02 to 0.03 mm; the volume density of joints, from 90 to 108 per meter; joint porosity from 0.114 to 0.206 percent; and the joint permeability from 15.6 to 31.7 md as shown in Table 1. Thus, the sample of core 28 (top) shows fairly effective reservoir capacity, but samples of core 26 (top) and core 27 (top) show poor effective reservoir capacity as shown in Figure 1.

The Paleocene Khairabad Limestone is gray to dark-gray, hard, compact, moderately to highly fractured, pelitomorphic to microcrystalline, commonly dolomitic, pyritic, and bituminous, fossiliferous (mainly foraminifers), somewhat argillaceous with abundant veins of recrystallized calcite. Microjoints are common and are filled either by calcite or bitumen. Open joints and voids also are present in certain beds. The average diameter of voids in sections is about 0.4 mm. Microjoints commonly are arranged chaotically, although horizontal, vertical, and inclined joints also are present. They show branches, some of them in the form of rough steps; others are curved; some are continuous to broken. In certain sections stylolitic sutures also are present. The average width of open or bitumen-filled joints ranges from 0.01 to 0.04 mm, volume density from 39 to 295 per meter, joint porosity from 0.050 to 0.750 percent, and joint permeability from nearly 1 to 205 md as shown in Table 1. Thus samples of core 23 (top), core 24, and core 25 (top) may be considered to have effective reservoir capabilities, samples of core 25 (bottom) to have fairly effective reservoir capabilities, and samples of core 23 (bottom) and core 25 (middle) to be ineffective reservoirs as shown in Figure 1.

SELECTED REFERENCES

Smekhov, Ye. M., ed., 1969, Metodika izucheniya treshchinovatosti gornykh porod i treshchinnykh kollectorov nefti i gaza (Methods for study of rock fissuring and fractured oil and gas reservoir rocks): Vses. Neft. Nauchno-Issled. Geol.-Razved. Inst. Trudy, no. 276, 129 p.
—— and M. M. Bulach, 1962, Handbook on methods for investigating jointiness of rocks and fissured oil and gas reservoirs, p. 68–77 (partly translated from Russian by S. Priadchenke for laboratory use only).